Roukaya Issaoui

Sustainable phosphate management

Environmental and Social life cycle assessment of phosphate mining in Tunisia

Sustainable phosphate management

Environmental and Social life cycle assessment of phosphate mining in Tunisia

by
Roukaya Issaoui

KIT Scientific Publishing

Karlsruher Institut für Technologie
Institut für Technikfolgenabschätzung und Systemanalyse (ITAS)

Sustainable phosphate management: Environmental and Social life cycle assessment of phosphate mining in Tunisia

Zur Erlangung des akademischen Grades einer Doktorin der Philosophie von der KIT-Fakultät für Geistes- und Sozialwissenschaften des Karlsruher Instituts für Technologie (KIT) genehmigte Dissertation

von Roukaya Issaoui

Tag der mündlichen Prüfung: 27. Juni 2022
Erster Gutachter: Prof. Dr. Armin Grunwald
Zweiter Gutachter: Prof. Dr. Jörg Woidasky

Impressum

KIT Scientific Publishing

Karlsruher Institut für Technologie (KIT)
KIT Scientific Publishing
Straße am Forum 2
D-76131 Karlsruhe

KIT Scientific Publishing is a registered trademark
of Karlsruhe Institute of Technology.
Reprint using the book cover is not allowed.

www.ksp.kit.edu

Print on Demand 2023 – Gedruckt auf FSC-zertifiziertem Papier

ISBN 978-3-7315-1248-6
DOI 10.5445/KSP/1000151757

Acknowledgment

I would like to express my sincere thanks to the following people, without whom I would not have been able to undertake this journey and complete this research.

My deepest gratitude goes to the Team of ENRES doctoral program, especially to my supervisors Prof. Dr. Armin Grunwald at the University of Karlsruhe Institute of Technology and Prof. Dr. Jörg Woidasky at the University of Applied Science of Pforzheim, not only for their administrative, financial and academic support but also for elevating the quality of my work to a high standard.

My sincere thanks to Pro. Dr. Mario Schmidt head of the ENRES program for giving me the opportunity to persue my dream and to the DAAD for financing the PhD. My special thanks to Prof. Dr. Tobias Viere at INEC, for bringing his considerable knowledge and experience to this thesis. His scientific knowledge allowed my PhD to go the extra mile.

My sincere appreciation to Dr. Christina Rösch at ITAS for bringing a strong academic input to my research.

Many thanks to the engineers and employees of the "Centre de Recherche" as well as all the CPG employees, for opening their doors to me and considering me part of them, without wheir colloaboration and hlep I would have no content for my thesis.
Thanks to my colleagues at ITAS and INEC, who have supported me and had to put up with my stresses and moans for the past three years of study!

I would like also to recognize the role of Dr. Tobias Dominik as my friend and my partner for his tremoudous scientic and human support; thank you very much without your critisime, motivation and support I would have stopped the PhD a long time ago.

Lastly, many thanks to my family in Tunisia for believing in me and for their infinit love, making the distance and the stress of a new life bearable.

Abstract

Phosphorus is a crucial element in agriculture to feed the fast-growing global population. A sustainable phosphorus supply has social and human dimensions as phosphorus serves a basic need, which is nutrition. Moreover, phosphate rock and white phosphorus are classified as critical to European industries, and the call to reduce dependency for importation through phosphorus recycling can mitigate the supply risk. However, phosphate rock can also be critical for producing emerging countries, as phosphate mining contributes positively to their national economies. Therefore, the question arises as to what will happen to emerging countries if the solution proposed by importing countries call to reduce the dependency on primary resources?

The hypothesis discussed is that the mining sector and the stability of the country are nurturing each other. This thesis investigates the implication of environmental and social impacts and resource governance of producing countries to mitigate global phosphate criticality. Tunisia was used as an example to investigate the implication of environmental and social impacts of phosphate mining

First, from a global level, the reserves as well as the supply risk of phosphate were investigated. The results show that currently, the global reserves are 73.1 billion metric tons. Phosphate rock reserves are not scarce as they are dynamic. However, the quality is decreasing, and the environmental impacts related to phosphate mining are increasing. Besides, the global supply risk factors are mainly the social and environmental impacts of phosphate rock mining in producing countries.

Second, to study the implication and the association of environmental and social impacts, the Environmental and Social Life Cycle Assessment (LCA) method was used in a local context. The studied area was the phosphate rock mining in the southeast region of Tunisia (the Gafsa region). In addition, a survey among the local population of the mining region was used to compare the results of the S-LCA with the perception of the local population.

The results of the E-LCA show that the environmental hotspots of phosphate rock mining are located in the mining region. The main impact categories are human health (243 thousand metric tons of NMVOCs per year), climate change (134 thousand metric tons of CO_2 eq per year), Human Toxicity (87 thousand metric tons of 1,4DB eq per year), PM10 emission (65 thousand metric tons of PM10 per year), and water depletion (12 million m^3 per year). The results of the S-LCA show that the phosphate industry is impacted by the national and regional socio-political situation in the country and the mining region. Reciprocally the phosphate industry is affecting the Tunisian economy and the regional socio-economic as well as the environmental situation in Gafsa. In comparison to the survey results, there are differences concerning the pressure on fresh water and the health status of the local population. Both fresh water consumption and the health of the local population are medium and low risk, respectively, in contradiction to the perception of the local population. Another problem of the region is the high local unemployment, which is double the national unemployment level. The reason for this is the high concentration of economic activity in the region and the non- diversification of economic opportunities.

We conclude that phosphate rock is also critical to producing countries as the health of their economy depends on the performance of the mining sector. Both the results of E-LCA and S-LCA show that the environmental and social impacts are correlated as the burden of adverse environmental impacts can lead to negative social change. Therefore, responsible mining is a cornerstone of mitigating the supply risk factors and ensuring a sustainable supply chain. Good phosphate mining practices could support local social stability.

Abstract German

Phosphor stellt ein grundlegendes Element in der Landwirtschaft dar, um die schnell wachsende Weltbevölkerung zu ernähren. Eine nachhaltige Phosphorversorgung hat damit eine soziale und eine humanitäre Dimension, da Phosphor mit der Ernährung einem menschlichen Grundbedürfnis dient. Außerdem werden Phosphatgestein und weißer Phosphor als kritisch für die europäische Industrie eingestuft und es besteht die Forderung, die Abhängigkeit von Importen durch Phosphorrecycling zu verringern, um das Versorgungsrisiko zu mindern. Phosphatgestein kann jedoch auch für produzierende Entwicklungsländer von kritischer Bedeutung sein, da der Phosphatabbau einen positiven Beitrag zu deren Volkswirtschaften leistet. Es stellt sich daher die Frage, was mit den Entwicklungsländern geschieht, wenn die von den Importländern vorgeschlagene Lösung einer Verringerung der Abhängigkeit von Primärressourcen umgesetzt wird.

Die diskutierte Hypothese lautet, dass der Bergbausektor und die Stabilität eines Landes sich gegenseitig begünstigen. In dieser Arbeit wird untersucht, welche Auswirkungen die ökologischen und sozialen Auswirkungen und die Ressourcenverwaltung der produzierenden Länder haben, um die globale Phosphatkritikalität zu mildern. Am Beispiel Tunesiens werden die Auswirkungen des Phosphatabbaus auf Umwelt und Gesellschaft untersucht.

Zunächst wurden auf globaler Ebene die Reserven und das Phosphatlieferrisiko analysiert. Die Ergebnisse zeigen, dass die weltweiten Reserven derzeit 73,1 Milliarden Tonnen betragen. Die Reserven an Phosphatgestein sind nicht als knapp zu bezeichnen, da sie sich dynamisch verhalten. Die Qualität nimmt jedoch mit zunehmender Erschließung ab und die mit dem Phosphatabbau verbundenen Umweltauswirkungen nehmen zu. Außerdem bestehen die Risikofaktoren für die globale Versorgung hauptsächlich in den sozialen und ökologischen Auswirkungen des Phosphatabbaus in den Produzentenländern.

Zweitens wurde zur Untersuchung der Auswirkungen und des Zusammenhangs von Umwelt- und Sozialauswirkungen die Methode der ökologischen und sozialen Lebenszyklusanalyse (LCA) in einem lokalen Kontext angewandt. Das untersuchte Gebiet für den Phosphatabbau liegt in der südöstlichen Region Tunesiens (Region Gafsa). Darüber hinaus wurde eine Umfrage unter der lokalen Bevölkerung in der Bergbauregion durchgeführt, um die Ergebnisse der S-LCA mit der Wahrnehmung der lokalen Bevölkerung zu vergleichen.

Die Ergebnisse der E-LCA zeigen, dass sich die Umwelt-Hotspots des Phosphatsteinabbaus innerhalb der Bergbauregion befinden. Die wichtigsten Auswirkungskategorien sind die menschliche Gesundheit (243 Tausend Tonnen NMVOCs pro Jahr), der Klimawandel (134 Tausend Tonnen CO_2-Äquivalente pro Jahr), die Humantoxizität (87 Tausend Tonnen 1,4DB-Äquivalente pro Jahr), die PM10-Emissionen (65 Tausend Tonnen PM10 pro Jahr) und die Wasserverschmutzung (12 Millionen m^3 pro Jahr). Die Ergebnisse der S-LCA zeigen, dass die Phosphatindustrie von der nationalen und regionalen sozio-politischen Situation im Land und in der Bergbauregion beeinflusst wird.

Umgekehrt wirkt sich die Phosphatindustrie auf die tunesische Wirtschaft und die regionale sozioökonomische Situation, sowie auf die Umweltsituation in Gafsa, aus. Im Vergleich dazu gibt es in den Umfrageergebnissen Unterschiede in Bezug auf die Bedeutung des Wassers und des Gesundheitszustands der lokalen Bevölkerung. Sowohl der Frischwasserverbrauch, als auch der Gesundheitszustand der lokalen Bevölkerung, sind im Gegensatz zur Wahrnehmung der lokalen Bevölkerung mit einem mittleren bzw. geringen Risiko behaftet. Ein weiteres Problem der Region ist die hohe lokale Arbeitslosigkeit, die doppelt so hoch ist wie die landesweite Arbeitslosigkeit. Der Grund dafür ist die hohe Konzentration der Wirtschaftstätigkeit in der Region und die mangelnde Diversifizierung der wirtschaftlichen Möglichkeiten.

Abschließend kann festgestellt werden, dass Phosphatgestein auch für die Erzeugerländer von entscheidender Bedeutung ist, da die Gesundheit ihrer Wirtschaft von der Leistung des Bergbausektors abhängt. Sowohl die Ergebnisse der E-LCA als auch der S-LCA zeigen, dass die ökologischen und sozialen Auswirkungen miteinander korrelieren, da die Belastung durch nachteilige Umweltauswirkungen zu negativen sozialen Veränderungen führen kann. Daher ist ein verantwortungsvoller Bergbau ein Eckpfeiler zur Abschwächung der Risikofaktoren bei der Versorgung und zur Gewährleistung einer nachhaltigen Versorgungskette. Gute Praktiken beim Phosphatabbau könnten die soziale Stabilität vor Ort fördern.

Table of Contents

List of Figures

List of Tables

List of Abbreviations

°C	Degree Celsius
1,4DB	1,4-dichlorobenzene
AAGR	Average Annual Growth
ANFO	Ammonium Nitrate Fuel Oil
Ba	Barium
benef. PR	Beneficiated Phosphate rock
Bn. t	Billion metric tons
CaO	Calcium Oxide
Cd	Cadmium
CO	Carbon Monoxide
CO_2	Carbon Dioxide
CPG	Company of Phosphates of Gafsa
CPRT	Crude Phosphate rock treated mechanically
Cr	Chromium
CRM	Critical raw materials
Cu	Copper
DAP	Diammonium phosphate
DERA	German Mineral Resources Agency
DNT	Tunisian Dinar
EC	European Commission
ECB	European Central Bank
EEA	European Environment Agency

E-LCA	Environmental Life Cycle Assessment
EPI	Environmental Performance Index
eq	Equivalent
EU	European Union
F	Fluorine
FTDES	Forum Tunisien des Droits Economiques et Sociaux, Forum Tunisien pour les Droits Economiques et Sociaux
g/l	Gram per liter
GCT	Groupe Chimique Tunisien
GDP	Gross domestic product
GHG	Greenhouse gases
GLO	Global
H_3PO_4	Phosphoric acid
HDI	Human Development Index
HHI	Herfindahl-Hirschman Index
IEF	Index of Economic Freedom
IHDI	Inequality-adjusted Human Development Index
IMF	International Monetary Fund
INS	Institut of National Statistics
INSP	Institut National de la Santé Publique
ISO	International Organization for Standardization
ITC	International Trade Center
kg	Kilogram
kg-km	Kilogram-kilometer
km^2	Square Kilometer

kWh	Kilowatt-hour
LCA	Life Cycle Assessment
LCI	Life Cycle Inventory
LCIA	Life Cycle Impact Assessment
m	Meter
$m.s^{-1}$	Meters per second
m^2	Square meter
m^2*a	Square meter annual
m^3	Cubic meter
MAP	Monoammonium phosphate
MgO	Magnesium Oxide
MJ	Megajoule
mm	Millimeter
Mm^3	Million cubic meters
MMT	Million Metric Tons
M	Molybdenum
N_2	Nitrogen
NACE rev.2	Nomenclature of Economic Activities revised version 2
Ni	Nickel
NMVOCs	Non-methane volatile organic compounds
NO_x	Nitrogen Oxides
NRC	National Research Council
NRGI	Natural Resource Governance Institute
OECD	Organisation for Economic Cooperation and Development

P	Phosphorus
P_2O_5	Phosphorus Pentoxide
P_4	White phosphorus
Pb	Lead
PCL_3	Phosphorus Trichloride
PM10	Particulate matter with a diameter of less than 10 µm
$Pm^2.5$	Particulate Matter with a diameter of 2.5 micrometers and smaller
ppm	Parts per million
PR	Phosphate rock
R/C	Resource to consumption ratio
RGI	Resources Governance Index
RoW	Rest of the World
S.R	Supply risk
Se	Selenium
SETAC	Society of Environmental Toxicology and Chemistry
S-LCA	Social Life Cycle Assessment
SO_2	Sulfur Dioxide
SOTEMU	Société tunisienne d'explosifs et de munitions
SO_x	Sulfur Oxides
Sr	Strontium
STTPM	Société Tunisienne de Transport des produits miniers
t	Metric ton
t/h	Ton/hour
Th	Thorium

tkm	Ton-kilometre
TN	Tunisia
TSP	Triple Super Phosphate, Total Suspended Particles
U	Uranium
UGTT	Union Générale des Travailleurs Tunisiens
UNEP	United Nations Environment Programme
US EPA	United States Environmental Protection Agency
USBM	United States Bureau of Mines
USD	US Dollar
USGS	United States Geological Survey
V	Vanadium
WGI	Worldwide Governance Indicator
WGI_{PSAV}	Worldwide Governance Indicator, Political stability and absence of violence
WHO	World Health Organization
yrs	Years
Zn	Zinc

1 General Introduction

1.1 Introduction

Phosphorus (P) is an essential element for all forms of life. It is the 11th most abundant element on earth. It is a non-renewable resource, and there is no substitute for it. It is classified as a macronutrient because living organisms require a large amount of phosphorus.

For plants, phosphorus is one of the three macronutrients added to the soil as a fertilizer (Nitrogen, Phosphorus, and Potassium). Phosphorus is not found by itself in elemental form. Phosphate rock (PR) is the primary source of phosphorus fertilizers, and it must be chemically treated to convert the phosphate into available forms for plant nutrition.

The global demand for nutrients and fertilizers is growing as fast as the world population's growth. As phosphate rock is a finite and non-substitutable resource, nature is not able to regenerate its stock as fast as our consumption rate.

In addition, following the current linear economic paradigm, the depletion of phosphate rock is a major threat to the ability to meet the nutrition needs of a fast-growing world population. Today's question is how we can continue to feed the world if there is not enough phosphorus to produce food. Therefore, the phosphorus criticality challenge is not only about the economic impact associated with a risk of phosphorus supply but also, and above all, it is crucial to meet a vital need for humanity: nutrition.

Moreover, due to the globalization of the economy and thus the supply chains, we need to address the management of raw materials as a global challenge in which its change impacts every economic player involved in the value chain. In the global market of raw materials, the risk of supply is not a matter only for the demand side (the buyers on the global market) but also for the supply side (countries who provide phosphorus to the global market) as these commodities represent high economic and social values for their countries. A sudden variability of the demand or the offer can create a distortion amplified through the supply chain with high economic consequences. According to Scholz et al. (2013), phosphorus is a demand-pull market mainly driven by the fast-growing population and its nutritional consumption behavior [176]. In this work, the global supply of phosphorus is assessed from the offer side. Phosphate rock and phosphorus, like many other raw materials, are subject to an unpredictable variability of the offer from the supplier caused by the socio-political instability in producing countries, restriction of exportation, reserve depletion, or other factors which would reduce the availability of these commodities in the global market.

1.2 Functionalities of phosphorus

The main primary source for producing phosphorus is apatite mineral phosphate rock (PR). There are two main families of geological phosphate rock formation or deposit: the igneous or magmatic deposit (existing in Russia, Canada, Finland, Brazil, and South Africa) and the marine sedimentary deposit (existing in China, USA, North Africa, the Middle East, and other regions). The first represents 1% of the global reserves [191] and is the basis for 12% of the world's phosphate production [14], while the second is the basis for 80% of the world's phosphate production[1] [215]. In most cases, the direct application of phosphate rock is not possible. Phosphate rock has to be treated chemically, reacting with sulfuric acid to produce phosphoric acid. The latter is an essential intermediate product that serves the manufacturing of many final products, mainly fertilizers. The agri-food system is by far the first global consumer of phosphorus, with 82% used for agricultural fertilization, 7% as animal feed additives, and 1¬2% as a food additive [177]. In addition, phosphorus is used in a myriad of industrial applications in the chemical industry, pharmaceuticals, the body care industry, the fuel industry, and recently the battery industry. It is also used as an antifreeze, as a flame-retardant, for metal coating, and in many other applications.

1.3 Research background: Phosphorus criticality in the literature

According to Nedelciu et al. (2019), due to the lack of reporting on the phosphate supply chain, there are four areas of weak and incomplete reporting and information: phosphate rock reserves and resources, losses along the supply chain, environmental and socio-political externalities of phosphate mining, and the access to data [142]. The authors stated

[1] A third minor source of phosphate is Guano phosphate. Guano applies to natural mineral deposits consisting of excrements, eggshells and carcasses of dead seabird and become rock-like during aging processes. Peru is the world largest producer of Guano with 30,000 tons are harvested in 2010 [133].

that to achieve global phosphorus governance and human rights for food, more reporting on pollution, international conflicts, and impacts on stakeholders is essential [142].

Considering the global phosphate supply chain as a black box [142] is the starting point of our research towards bringing insights into the current global phosphate supply chain. This section starts with a literature review on the concept of phosphorus criticality, then the development of the hypothesis, and finally, the research roadmap and structure of the thesis.

1.3.1 Critical mineral definition

According to the American National Research Council (NRC), the term "critical mineral" appeared for the first time in the Strategic and Critical stockpiling Act of 1939 for the USA [141]. The criticality of minerals was introduced in the context of WWII and postwar reconstruction as a strategic mineral for national security.

With the accelerated economic and technical advances, the sophistication of products, and the globalization of markets, nations are competing to protect their competitive advantage by securing a reliable supply of minerals to meet in time the need of fast technological innovation. Moreover, the current economic paradigm, known as linear economy, is posing uncertainty on the availability of minerals in the long term to meet the growing need of the global population and future generations.

In this context, critical minerals are defined as minerals for which a threat to supply from abroad could involve harm to the nation's economy (Evans, 1993, cited in the National Research Council, 2008) [141]. The National Research Council (NRC) defined the critical mineral as follows: "To be critical, a mineral must be both essential in use and subject to supply restriction" [141].

The ad-hoc Working Group of the Raw Materials Supply Group chaired by the European Commission (EC) defined a critical mineral as: "A raw material is labeled "critical" when the risks of supply shortage and their impacts on the economy are higher compared with most of the other raw materials" [70].

In the definitions given above, there are two dimensions to define a critical material, which are the national economic importance of the material or national economic vulnerability and a high risk of a foreign supply shortage.

1.3.2 Literature review of phosphorus criticality assessment

Recently, various studies have been conducted to assess the supply risks of different raw materials and the vulnerability of national economies to a potential shortage of supply (see the full list of raw materials criticality studies used as a sample for the literature review in the Annex, Table A-1). For the literature review, 13 out of 23 studies were selected (see Table 1.1). The criterion for the selection was the assessment of phosphorus criticality.

Table 1.1 presents the method and indicators used in each criticality assessment study. There is no standardized method to assess the criticality of minerals. The method can change according to the scope of the assessment, which could be either national/regional or sector scope, and the time frame, which could be short or long term [2]. In national reports, criticality is defined by the importance of the mineral to the domestic economy and the risk of an external supply restriction. The critical minerals are those involved in mega sectors with a high contribution to the GDP of the country [72].

From the conducted literature review, the main indicators used to assess the supply risk dimension of criticality are:

- Country concentration: used in 12 studies out of 13
- Country socio-political instability risk: used in 11 studies out of 13

In addition to the methodological aspect of phosphorus criticality, the literature review shows that phosphorus criticality has mainly been investigated in the European Union [69, 73] and on a country or regional level, such as in Germany for the German economy [45][45] [4, 30] and the Bavarian federation in Germany [216–219]. This is explained by the fact that the European Union is dependent on phosphate rock and phosphorus importation. The European Union (EU) does not possess phosphate rock reserves except in Finland. Therefore, according to the European commission studies [19, 69, 70, 72], phosphate rock and white phosphorus are both critical for European industries and emerging technologies.

From the European perspective, the criticality of phosphate is justified by the high geo-concentration of phosphate rock reserves in countries with a high potential of socio-political instability and in China, where trade restrictions are applied [72]. In Germany, phosphorus was not considered critical in the past. It used to be classified in the medium-risk group of materials [42, 43], but since 2019, according to the German Mineral Resources

Agency (DERA), phosphate rock is considered a high-risk mineral and commodity for the German economy [44].

Even though a given mineral can be involved in different applications along its life cycle, the risk of supply and its consequences may have not the same degree for all applications. Thus, Gantner (2016) has assessed phosphorus criticality on the level of the global value chain [82]. It suggests not to aggregate the criticality of materials under economic importance and supply risk for a country or a region-level, but rather to assess the criticality of different functionalities of the material along the global value and supply chain. This method gives a holistic view of the risk of supply and its impact along the value chain with more information about which application (or sector) is more vulnerable to the change of supply and how they affect each other.

Table 1.1: Phosphorus criticality assessment in the literature

Title	Scope	Indicators	Phosphorus criticality
Rohstoffsituation Bayern: keine Zukunft ohne Rohstoffe (2009) (2011) [216, 217] (Raw material situation in Bavaria: no future without raw materials)	Bayern	**Quantitative indicators:** - Static lifetime - Global Politic Risk Index - Concentration in 3 countries - Concentration in 3 companies **Qualitative indicators:** - Importance for future technologies - Strategical and political use - Substitutability	Phosphate: High-risk group but low importance for the Bavarian industry
Kritische Rohstoffe für Deutschland (2011) [67] (Critical Materials for Germany)	Germany	**Vulnerability**: - German share of the world use - Sensitivity of the value chain - Emerging technologies impact on demand - Substitutability **Supply risk:** - Country risk of German imports - Country risk of global production - Country concentration of global reserves - Companies concentration in the global market - Static depletion time - By product ratio - Recycling	Phosphate belongs to the low criticality group: low supply risk, low vulnerability for German industry

Angebotskonzentration bei Metallen und Industriemineralen potenzielle Preis- und Lieferrisiken (2012) [42] (Supply concentration of metals and minerals, potential price and supply risks)	Germany	- Country concentration (HHI) - Country risk index - Share of the 3 larger countries	Phosphate: Medium risks group for the mining
Rohstoffsituation der bayerischen Wirtschaft (2015) (2017) [218, 219] (Situation of raw materials for the Bavarian economy)	Bayern	**Quantitative indicators:** - Static lifetime - Global Politic Risk Index - Concentration in 3 countries - Concentration in 3 companies **Qualitative indicators:** - Importance for future technologies - Strategical and political use - Substitutability	Phosphate: medium-risk group and medium importance for the Bavarian industry
Angebotskonzentration bei Metallen und Industriemineralen – Potenzielle Preis- und Lieferrisiken (2014) [43] (Supply concentration of metals and minerals, potential price and supply risks)	Germany	- Country concentration (HHI) - Company concentration (HHI) - Weighted country risk index - Share of the 3 larger countries - Net exports	Phosphate: medium-risk group for the mining sector high risk for traded commodities
Report on Critical Raw materials for the EU (2010) (2014) [69, 70]	European Union	**Economic importance**: - The share in end-use - EU Mega sector value **Risk of Supply:** - Substitutability index - Recycling input rate - Country concentration (HHI) - Country risk (WGI)	Phosphate rock is critical for the European economy
Angebotskonzentration bei Metallen und Industriemineralen – Potenzielle Preis- und Lieferrisiken: Liste DERA (2016) [44] (Supply concentration of metals and minerals, potential price and supply risks)	Germany	- Country concentration (HHI) - Company concentration (HHI) - Weighted country risk index - Share of the 3 larger countries - Net exports	- High risk group for mining - High risk for traded commodities

Ressourcenstrategische Betrachtung der Kritikalität von Phosphor (2016) [82] (Resource strategic consideration of the criticality of phosphorus)	Global value chain	- Phosphate rock - Phosphoric acid - White phosphorus - Derivate - Recycled Phosphorus	- Low and Medium criticality for PR - High criticality for White phosphorus and his derivates - Medium criticality for recycled Phosphorus
Study on the review of the list of critical materials (2017) [72] [19]	European Union	**Economic importance:** - The share of end-use in sectors (at a NACE rev.2) - The sector added value (at a NACE rev.2) - The substitution index related to the economic importance **Risk of Supply**: - Import reliance - Country concentration (HHI) - Country risk (WGI) - Trade restriction - Recycling input rate - Substitution index related to supply risk	Both phosphate rock and white phosphorus are critical

1.4 Hypothesis

Criticality of minerals has two dimensions: (1) the importance of the mineral to the national economy, and (2) the risk of a foreign supply shortage. The economic importance takes the interest of the national economies and the short-term future technological development of importing countries into consideration. The supply risk is expressed mainly by the socio-political instability of countries producing those minerals, which in the case of phosphate, are mostly emerging countries.

Criticality of minerals was assessed from a national concern. The concern is centered on the reliability of foreign supplies. According to the Raw Materials Initiative of the European Commission, securing a reliable and sustainable supply of natural resources is crucial to the competitiveness and growth of the EU economy and the objectives of the Europe

2020 strategy [69]. The socio-political stability of the producing countries is the main element to calculate the supply risk.

In the case of phosphorus, the risk of the supply interruption is not only associated with national economic consequences but also with global consequences, as it is an essential element to sustain life and feed the global population. The need to ensure a global sustainable supply of phosphorus is above the economic importance of the material; it has humanitarian importance. Moreover, in a global market's conditions and according to the sustainable development goals (SDGs), national economic success cannot be entirely independent of international and global conditions. According to Grunwald (2010), a global perspective is a criterion to achieve sustainability, both in terms of conservation and the creation of minimum conditions for a decent life [91].

Another issue that has not been taken into consideration by current criticality assessment studies is the fact that the stability of countries, included in the equation to assess the supply risk, relies in many cases on the success of the mining sector. In other words, sustainable mining might be connected to the social stability of the country. Phosphate rock can also be critical to producing countries, as the contribution of the mining sector to the national economy usually is very high. Their economic dependencies imply the following questions: What will happen to emerging countries if the solution proposed by importing countries calls to reduce the dependency on primary resources? What will happen to communities living on the mining sector if they are excluded from solution development to reduce the criticality of minerals on a global level? Can we talk about the sustainable supply of phosphate without mitigating the social instability in the producing countries?

The hypothesis discussed here is that the mining sector and the stability of the country nurture each other. This thesis investigates the implication of environmental impacts and social impacts of the mining sector in producing countries to mitigate phosphate criticality. Tunisia, one of the global leaders of phosphate rock production, is an example to investigate the implication of environmental and social impacts of phosphate mining. The choice is based on the history of the mining region and the applied open-pit mining technique, which is widely used in the world for the production of sedimentary phosphate rock [215]. The Tunisian case can inspire other studies. However, this work does not claim that the Tunisian scenario would be the same in every phosphate mining region due to the differences in the cultural, social, and political context.

1.5 Research roadmap

The adopted approach is to look first into the claim that phosphate is a critical material and to investigate the status of the world's reserves. As pointed out before, from a European perspective, phosphate is critical because it is highly geo-concentrated in regions with high political and social instability. As such, Tunisia serves as an excellent example of a phosphate producing country. The phosphate mining region Gafsa knew a significant social movement in autumn 2010, which led to a socio-political revolution in the country, as well as a sharp drop in the phosphate production. The case of Tunisian phosphate mining, therefore, gives insights to understand the dynamic interaction between environmental and social impacts and how such interactions may affect access to natural resources.

Figure 1.1 presents the structure of this work. After the methodology is described in chapter 2, chapters 3 and 4 carry out a detailed analysis of the availability and criticality of phosphate rock from a global level. In chapter 5, the research scope is narrowed down to a regional level, and the case study of Tunisia is introduced. The chapters 6 and 7 present the results of the environmental life cycle assessment (E-LCA) and the social life cycle assessment (S-LCA) of the phosphate rock mining region in the southeast of Tunisia (Gafsa region). Finally, chapter 8 discusses the finding of the thesis related to the initial hypothesis, outlining the potential correlation of environmental and social impacts and the role of socio-environmental impacts of phosphate mining in the producing countries to mitigate the global supply risk of phosphate rock.

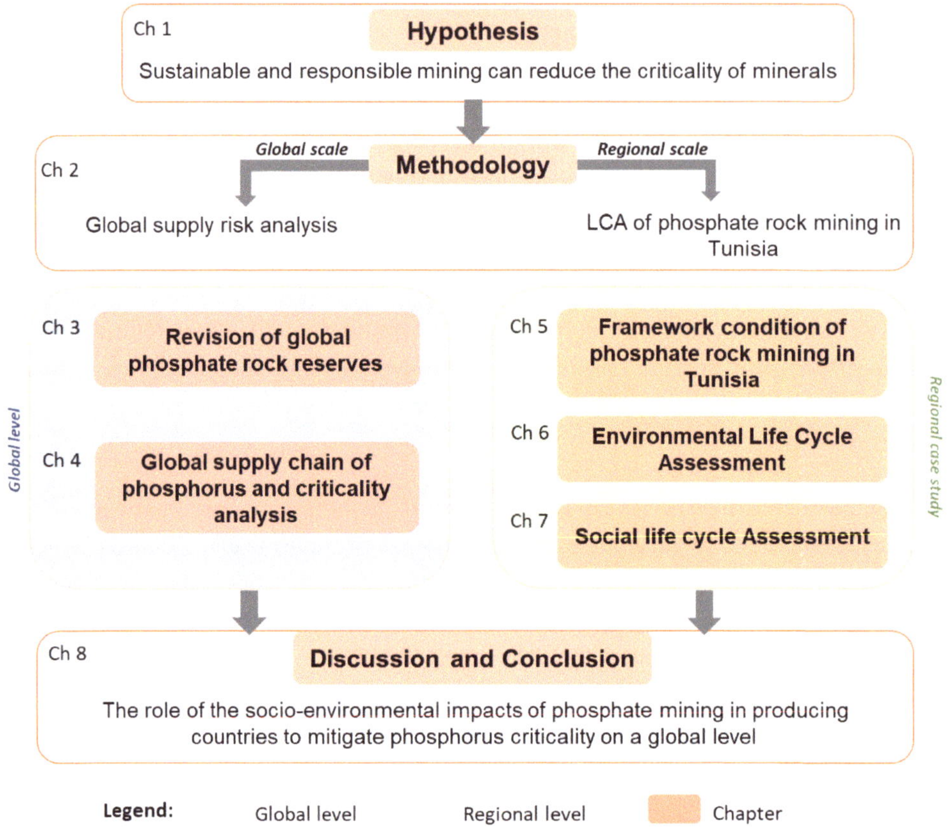

Figure 1.1: Structure of the document and roadmap of the hypothesis analysis

2 Methodology

2.1 Introduction

Mining is one of the oldest industries, and it was for a long time the motor of technological innovation. Human civilization is linked closely to the access to and the use of natural resources. The wealth of a nation was defined by its ability to control natural resources. Therefore, for long centuries, the competition to secure access to resources was a motive for wars, slavery, colonization, and human settlement in extraction areas. Nowadays, in an era of increasing scarcity of natural resources and the rise of public awareness about the climate situation and human rights around the world, competition among the world's major economies has shifted towards the development of more efficient and sustainable technologies to meet their ever-increasing need for minerals and metals.

The cornerstone of the thesis is to establish a link between the local and regional socio-environmental impacts of phosphate rock mining and the global supply risk of phosphorus. The analysis requires looking at the problem from different perspectives, from the importer's perspective, from the producer's perspective, and at the dynamic of the global market. It additionally requires different levels of analysis, such as the industry level, country level, and the global market level, to provide a better problem understanding and allow realistic and sustainable phosphorus management.

2.2 General Approach

For this work, the applied approach combines two levels of analysis. The supply risk factors were investigated on a global supply chain level and a regional production level. On the regional level, the analysis of the sustainability of the mining industry is multi-dimensional (environmental and social).

Figure 2.1 shows that the supply risk assessment of phosphate is dynamic. Many factors can mitigate or trigger a supply risk. For instance, technical innovation may be an answer to the physical availability of minerals and increase access to previous non-economically or technically feasible resources. For the risk of trade restrictions or embargos on the import of conflict minerals, it is the domain of international relations and politics and can be mitigated through trade agreements and foreign investments in the mining sectors in emerging countries. For the socio-political factors, they are tightly bound to the mining activities and the conditions in the producing country. Therefore, it is useful to start the analysis from the local context of mining to judge the severity of the supply risk on the global market and to adapt decisions according to it.

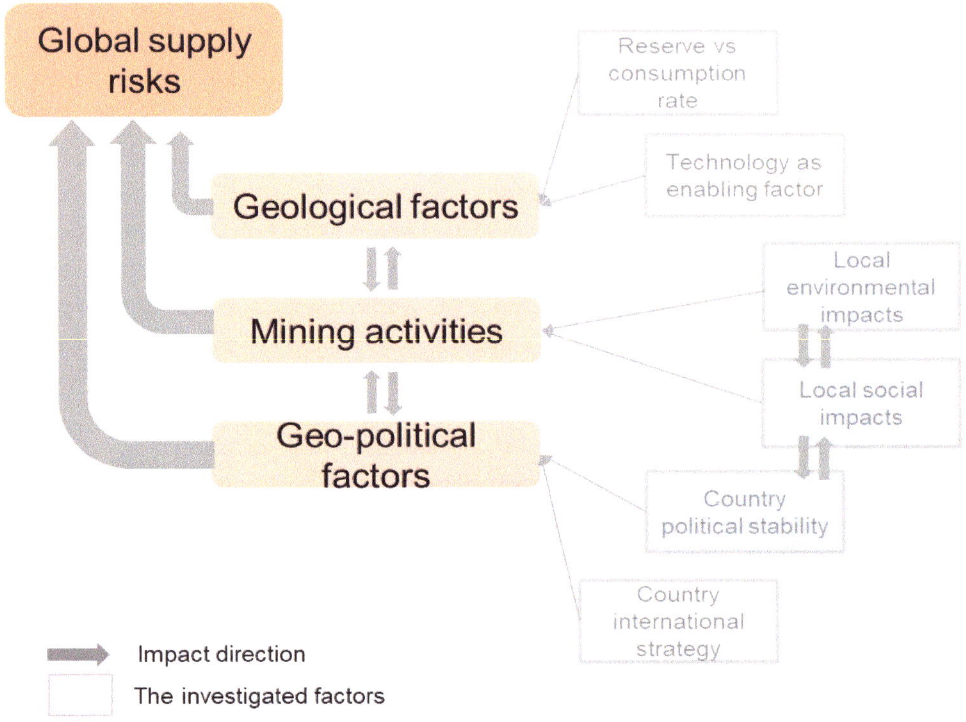

Figure 2.1: General approach: the assessment of local environmental and social impacts of phosphate mining to mitigate the supply risk on the global market

2.3 Phosphate supply risk assessment on a global level

A set of parameters was used to identify factors that would increase the supply risk of the following key-intermediate phosphate products [82]:

- Beneficiated phosphate rock
- Phosphoric acid
- White phosphorus

Four parameters were used to assess the supply risk on a global level: (1) the geo-concentration, (2) the socio-political stability of the country, (3) the governance of natural resources - only applied the beneficiated phosphate rock as a natural resource - and (4) the environmental performance of the country. The parameters and indicators are shown in Table 2.1. The data reference year is 2017.

The indicators were selected based on the results of the literature review. The parameter 'geo-concentration' is measured with the indicator 'country concentration' expressed by the 'Herfindahl Hirschman Index' (HHI).

The parameter 'socio-political stability' is measured with the indicator 'political stability and absence of violence' expressed by the 'Worldwide Governance Indicator[1]' (WGI_{PSAV}). It reflects the potential of supply interruption due to social and political problems in a country.

In addition to parameters from the literature review, two other parameters were used as they were country and sector-specific. The two parameters were 'Resource governance' and 'Environmental performance'. They reflected the production conditions in producing countries, which might be implicated in the supply risk. The parameter 'Environmental performance' was introduced by Graedel et al. [87]. It indicates the potential of environmental implications using a particular mineral [87].

The 'resource governance' parameter was measured with the indicator 'Resource Governance Index' (RGI), which is specific to the extractive sector of a country. It was used to investigate how phosphate is managed in a country as natural resources.

The 'Environmental performance' parameter was measured with the 'Environmental Performance Index' (EPI), which expresses the pollution level of the country.

[1] The Worldwide Governance Indicators (WGI) construct aggregate indicators of six broad dimensions of governance:
Voice and Accountability/Political Stability and Absence of Violence/Government Effectiveness/Regulatory Quality/Rule of Law/Control of Corruption [223]. Only the WGI of Political Stability and absence of violence was used for this study. it is referred to as WGI_{PSAV}

Table 2.1: Parameters and indicators to express the supply risk of Phosphate rock and Phosphorus

Parameters	Indicator	Measure	Source of Data
Geo-concentration	Country concentration (HHI)	0 < 0.1 low concentration; from 0.1 - 0.18 moderate concentration; > 0.18 high concentration	USGS [212]
Socio-political stability	Political stability and absence of violence (WGI_{PSAV})	ranges from -2.5 (weak) to 2.5 (strong)	World Bank[223]
Resources Governance	Resource governance Index (RGI)	Ranges from: 0 - 100: <= 30: failing, >= 75: good	Resource Governance Institute[144]
Environmental performance	Environmental Performance Index (EPI)	Ranges from: 0 (failing) to 100 (good)	Yale Center for Environmental Law & Policy [222]

2.3.1 Supply risk indicators definition

- **Market concentration : HHI**

The HHI is an indicator to calculate market concentration. Its formula is:

$$HHI = \sum_{i=1}^{n} (country's\ production\ share)_i^{\ 2}$$

n: the number of countries in the global market.

The country's production share is expressed as a percentage. If the value of the production share is a decimal fraction (e.g., 30% as 0.30), then the HHI ranges from 1/n to 1. If the value of HHI is very close to zero, it indicates perfect competition. If the value is one, it indicates a monopoly.

The horizon merger guidelines of the U.S. Department of Justice and the Federal Trade Commission considered an HHI between 0.10 and 0.18 to be moderately concentrated and indices above 0.18 to be concentrated [195]. However, in 2010 they issued an update of the merger guidelines, now considering an HHI above 0.25 as a highly concentrated market [196]. According to Achzt and Helbig (2013), the German Federal Ministry of Economics and Technology considers a material with an HHI above 0.15 as concentrated [2]. The working paper on the European market concentration for the European Central Bank (ECB) used the HHI > 0.18 as a threshold for a high level of market concentration [129].

For this work, the threshold for the HHI of 0.18 is used to consider a market highly concentrated. The classification of the market concentration is as followed:

- Non-concentrated market: HHI < 0.10
- Moderately concentrated market:0.10 < HHI <0.18
- Highly concentrated market: HHI above 0.18

This indicator shows only how much the market is concentrated, but it does not express the vulnerability of access to the sources. In other words, the country concentration indicator might increase the criticality of the material if the country or the company that controls the production is considered risky in terms of socio-political stability.

- **Political stability absence of violence**

The indicator 'Political Stability and Absence of Violence' is one of the six indicators of the World Governance Index (WGI) [118]. The indicator measures the likelihood of political instability and/or politically-motivated violence. It ranges from -2.5 to 2.5, with a low score indicating a higher political risk in the country [118].

- **Resource Governance Index: RGI**

The RGI measures the existence of policies and the applicability of rules and laws in a country related to the management of natural resources, such as extraction permits, revenue management, management of impacts, and enabling factors like corruption control, the rules of law, and political stability of the country [144].

- **Environmental performance index: EPI**

The EPI is a composite index that captures the capability of a country to establish environmental policy goals. It assesses countries based on 24 performance indicators across ten issue categories, covering environmental health (air quality, water sanitation, and heavy metals) and ecosystem vitality (agriculture, water resources, air pollution, climate and energy, fisheries, biodiversity, and forests) [222].

2.4 Environmental and Social Life cycle assessment of phosphate mining in Tunisia

According to the International Reference Life Cycle Data System handbook (ILCD) (2010), "Life Cycle Assessment (LCA) is a structured, comprehensive and internationally standardized method" [74]. For the environmental impacts, the LCA is part of the ISO 14040 and 14044 standards [114]. Social Life Cycle Assessment (S-LCA) is more recent than environmental (E-LCA) and economic impacts (LCC). It comes to complete the full image of the sustainability assessment of supply chains of products [4]. A framework for Social LCA has been adapted from the ISO standardized E-LCA Framework as outlined in the guidelines on SLCA of UNEP/SETA [140, 202].

The LCA is a technique to identify, quantify, and assess the environmental (E-LCA) and social (S-LCA) impacts related to the product's life cycle from the raw material extraction to its disposal at the end of life, including the transformation and use phase [74, 202].

The E-LCA and S-LCA have the same structural framework of ISO. Together they bring more completion to the sustainability assessment of the product and its value chain. The difference is the focus. While the E-LCA focuses on the product system and the physical flow specific to a site or facility and the process used at the production site, the S-LCA focuses on the enterprise management level using data reported at the enterprise, such as labor practice [202]

For the mining sector, the use of the LCA technique aims to integrate the environmental impacts of mining technologies and processes with social impacts related to the management of natural resources and its social environment to provide a comprehensive way of assessing the sustainability of mining practices.

For this work, both E-LCA and S-LCA have been conducted using the openLCA free, open-source software. The database and the assessment methods are shown in Table 2.3. The details of the assessment methods can be found in Annex B.

Table 2.2: Comparison of E-LCA and S-LCA: common features, differences, and complementaritie

Common features	Differences	Complementarities
▪ Common ISO framework: Goal, Scope, Life cycle Inventory, impact assessment, and interpretation ▪ Iterative procedure ▪ Identification of hotspots ▪ Expression of impacts related to a functional unit when the data used is quantitative ▪ Provides useful information for decision-maker	▪ The focus of E-LCA is to evaluate the environmental impacts using data on physical quantities related to the product (production) ▪ The S-LCA focus on the assessment of socio-economic impact using data on organization related aspect (management)	▪ A more comprehensive picture of the product's life cycle to make decisions adopting a sustainability perspective ▪ Reveal the possibility that the burden of one dimension of sustainability would shift to another dimension [48]

Table 2.3: Method and tools used to conduct the E-LCA and the S-LCA of beneficiated phosphate rock production in the mining region of Gafsa

	E-LCA	S-LCA
Scope of the LCA	From Cradle to Gate	From Cradle to Gate
Modeling software	OpenLCA version 1.10.2	OpenLCA version 1.10.2
Database	Ecoinvent version 3.4	PSILCA version 1.7
Impact assessment method	ReCiPe Midpoint (H) v1.11	Social Impacts Weighting method

2.4.1 Stakeholder categories definition for S-LCA

Stakeholders are people, groups, or institutions that are likely to be affected by a proposed intervention (either negatively or positively) or those which can affect the outcome of the intervention [168].

First, a stakeholder classification list was developed based on the answer to the questions adapted from the work of the World Bank on social assessment tools and techniques [168]. The questions are:

- Who are the people/groups that have an interest in phosphate mining? What is their interest?
- Which stakeholder is involved in the creation of value? What is their role?
- Who might be directly and indirectly impacted by the phosphate mining activities?
- Who has the power to influence the company's work (positively or negatively)?

- **EPI Classification of stakeholders: Power/Interest matrix**

The stakeholder power interest grid is a management tool to classify the company's stakeholders into four categories according to their level of influence on the company's work and their interest in it.

Some stakeholder groups may have the power either to block the company's work or to facilitate it. Some may be interested in the value created by the company and its process, while others may not be interested. A list of stakeholders was sent to the research center of the company (see Annex B, Table B-5 and Table B-6), in which participants listed stakeholders of the CPG and attributed a degree of relevance (power/interest). The stakeholders were prioritized according to their degree of interest in the company's work and their influence on it.

- **EPI Classification of stakeholders: Power/Interest matrix**

According to the participants' answers to the stakeholders' Power/Interest Matrix, stakeholders who have high power and interest were selected as the stakeholder category for the S-LCA study. Those stakeholders are directly affected by and/or have a direct effect on the mining activities (see Annex B, Figure B-2).

2.4.2 Qualitative assessment of social impacts

The S-LCA provides insights into the social conditions of production, use, and disposal for decision-makers [202]. Nevertheless, an S-LCA is not enough to develop decisions and actions that consider the socio-economic key issues specific to the region. According to Grunwald (2010), the impact assessment of technology has to be participative, and the 'local knowledge,' which is unknown to the 'distant' experts and decision-makers, should be used to support decision-making. This is particularly relevant in the case of local and regional problems [91].

- **A Survey among the local population**

The social conditions of production would be assessed by seizing the perception of directly-affected stakeholders. Thus, in addition to the S-LCA, a site-specific social assessment was conducted using a qualitative survey.

At a first attempt, the directly-affected stakeholders who were chosen for the survey are the employees and the local population. Due to the social instability in the company (CPG), the perception survey with employees was not possible, and the on-street perception survey with the local population was not easy to conduct. Therefore, an online survey was designed to complement the street survey among the local population.

For the stakeholder category local population, a survey was developed to collect data about the social impacts: health, economic opportunity, water depletion and pollution, and environmental load. These social impacts are expressed according to the perception of the local population and to be compared to the results of the social life cycle inventory assessment (S-LCIA).

After defining the social impacts, the indicators were defined to guide the questionnaire development (see Annex B, Table B-6). The survey is composed of 10 ordinal scale questions. The average duration to answer the questions is 7 minutes, (survey questions: see Annex B, Table B-7).

At first, a pre-test was conducted with five people. The goal was to test the questionnaire's clearness and its ability to provide interpretable answers in French for the online survey and Arabic for the on-street survey. The second step, the target population, was defined as well as the channels to reach the local population (see Annex B, Figure B-3). The target population is the population living in the mining basin and where the phosphate industry is concentrated (South district of Gafsa, Metlaoui, Moulares, Redeyef, and M'dhila). The diversification of the channels of communication helped to maintain a diversification of the studied sample according to the cultural context of the studied area. The channels used are coffee shop (the meeting point of men older than 15 years, mostly unemployed young men and retired men), bus station (active population in their way to work), the weekly market (Souk in Arabic), which is a very important meeting point of all the mining villages where women are very present. The online survey was developed using the Survey-Monkey website [187]. The channels are social media regional groups and pages on Facebook and by emailing the link.

After defining the sample, the questions, and the communication channels, a street survey, with the local population was conducted for three weeks in April 2019. However, the access to the online survey was open for two months (Mid April to Mid June, 2019).

3 Revision of the global phosphate rock reserves

3.1 Introduction

As a finite resource, phosphate rock was subject to many studies investigating its availability and criticality [39, 41, 82, 175]. This chapter presents an updated inventory of the Tunisian phosphate reserves, which leads to an update of the global reserves of phosphate rock according to the latest data available from the USGS 2019 and the Company of Phosphates of Gafsa (CPG) 2018 in Tunisia [213][213] [136]. In addition, the geo-concentration of global phosphate rock reserves, as well as the global trade of phosphate rock, will be discussed. The geo-concentration of phosphate rock is a parameter to measure the criticality of phosphorus, a concept that will be discussed in the next chapter. The global trade of phosphate rock will give insights about the global supply chain of phosphate, the geo-concentration, and the evolution of demand.

3.2 Definition and report terminology

To avoid any ambiguity using different terms to describe the occurrence of mineral resources, the reporting terminology for this work uses the mineral resources classification system of the United States Bureau of Mines (USBM), the United States Geological Survey (USGS) [194], and the definition of the European Commission [70]. Figure 3.1 displays a simplified version of the classification system of the mineral resources and reserves.

3.2.1 Resources

Natural resources are the stock of renewable and nonrenewable materials, to which human societies attribute a particular value and use them to increase their well-being. They can be land, food, water, energy (solar, wind, oil, gas), and minerals (metallic and non-metallic minerals). Resources are defined and classified according to the domain of interest. For example, the extractive sector classification system of mineral resources is not the same as the water resources classification system in the water management sector.

The USGS geological survey circular 138 defines mineral resources as follow:

*"**Resource** – A concentration of naturally accruing solid, liquid, or gaseous material in or on Earth's crust in such form and amount that identified as a potential commodity, which would be economically feasible to be extracted and exploited in the future" [194].*

The European Commission (EC) in their report on critical materials defines resources as:

*"**Mineral resource** regroups all identified resources. It is a natural concentration of minerals or a body of rock that is, or may become, of potential economic interest as a basis for the extraction of a mineral commodity. A resource has physical and/or chemical properties that make it suitable for specific uses, and it is present in sufficient quantity to be of intrinsic economic interest" [70].*

According to Figure 3.1, the classification of mineral resources depends on the degree of geological, technical, and economic probability or evidence. For instance, the discovered resources are classified as identified resources when their location, grade, quality, and quantity are estimated according to geological evidence. The degree of geological evidence's confidence varies from inferred to measured resources. The inferred resources are those estimated according to a geological assumption based on geo-scientific data but not supported by sampling and measurement. The indicated resources are more probable than the inferred resources. Their quality and grade results are computed from measurement and sampling, but with a lower degree of assurance than the measured resources. The measured resources are those whose quality and grade computed results are proven through detailed sampling and give more accurate geological and physico-chemical properties such as size, shape, depth, and mineral content. The term "Resources" will be used in this document to refer to the identified resources, as indicated in Figure 3.1.

3.2.2 Reserves

According to the mineral resources classification system of the USGS, the Reserves Base is equal to the sum of indicated and measured resources represented as the demonstrated resources in Figure 3.1. The demonstrated resources can be considered as reserves base if they can be mined and produced according to the current technical practices, and they have good potential to be economically available.

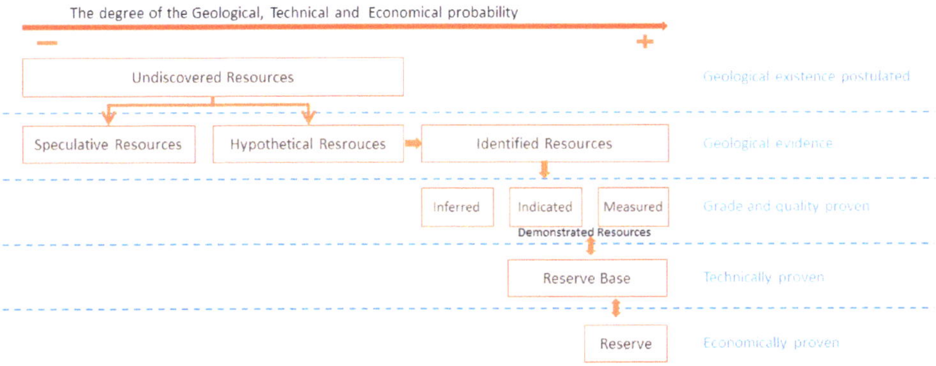

Figure 3.1: Classification of mineral resources occurrence according to their geological, technical, and economic evidence, based on the definition of the USGS [194] and the EC [70]

The reserves base can be classified related to the degree of their economic potential or evidence into (1) **Sub-economic Reserves** which cannot satisfy the current economic criteria, (2) **Marginal Reserves** which have a weak economic potential at the time of determination but can be commercially viable in the future due to economic and technological change, and (3) **Reserves** which meet the current mining and production economic criteria at the time of determination.

For this report, the term “Reserves” will be used to refer to the part, which has been geologically available in quantity and quality, and which is technically, economically, and legally proven to be mined and produced at the time of determination.

3.2.3 Other relevant definitions of mining terms

Following are definitions of the commonly used terms for mining:

Minerals are naturally occurring inorganic elements or compounds having an orderly internal structure and characteristic chemical composition, crystal form, and physical properties. Minerals generally form crystals and have specific physical and chemical properties, which can be used to identify them [199]. Minerals can be metallic and non-metallic ore. In this document, minerals will be used to refer to non-metallic ore such as phosphorus.

An **Ore** is a natural aggregation of one or more solid minerals that can be mined, processed, and sold at a profit [100].

A **Mining site** is a natural deposit of mineral or organic substances, whether exploited in an open-pit or underground according to the official definition of the mining code of Tunisia under the law n° 2003-30 [34].

Phosphate rock (PR) is the primary source of phosphorus. It is a common term to define unprocessed rock and beneficiated concentrate. The unprocessed phosphate rock contains different forms of phosphate minerals in a gangue of other minerals. The apatite form is the group of Calcium Phosphate minerals with the formula Ca5 (PO4)3(F, Cl, OH) [215]. The beneficiated phosphate concentrate is the product obtained after mining and processing phosphate ore.

For this document, the term "phosphate rock" will be used to indicate the phosphate ore that is the crude phosphate rock available to be mined and processed at a profit. The term "beneficiated phosphate rock" will denote the marketable phosphate rock with a concentration of 29% P_2O_5 and more.

Phosphorus Pentoxide (P_2O_5) is a chemical compound, which is commonly used as a denominator to express the phosphorus content of phosphate ore and the marketable beneficiated phosphate rock.

Phosphorus (P) is a non-metallic- chemical element; its atomic number is 15. Phosphorus has several forms or allotropes. The main forms of phosphorus used in the industry are white phosphorus and red phosphorus. Phosphorus is not substitutable for plant growth, thus for the fertilizer industry and agriculture.

Resource to consumption ratio or **R/C** is the estimated depletion time of the identified resources. The depletion time refers to the static lifetime of the identified resources. It considers that the consumption rate is constant while there are no newly available resources. The ratio is expressed in years [175]. In this document, the depletion time or R/C was calculated based on the reserves base's data (demonstrated resources) and the reserves.

3.3 Geographical distribution of Phosphate reserve

3.3.1 Previous global phosphate reserves according to the USGS database

According to the recent data from the United States Geology Survey (USGS) 2018, the global reserves of phosphate rock are estimated to 69.9 Billion tons in 2017 [212]. The sizes of the phosphate rock reserves, which are mainly concentrated in ten countries, are presented in Table 3.1. The share of the top 10 countries accounts for 94% of the reserves. The current R/C, using the USGS data for 2017, is 260 years. In 2010, just after updating the global reserves, the R/C was 360 years according to the USGS data [205, 212].

In 2011, the USGS published an update of Moroccan/Western Sahara and Syrian reserves, which increased the global reserves from 15.6 billion metric tons in 2009 to 65 billion metric tons [205] (see Table 3.1). They projected that the global phosphate rock reserves would increase as well as the mining production capacity due to new reserves in Saudi Arabia and an increase of the bulk capacity in Egypt and Jordan in 2017.

Table 3.1: Geo concentration of the global reserves in the top 10 countries expressed in thousand metric ton, according to the USGS database 2010 [205] and 2018 [212]

[10^3 metric tons]	2009	2017
Morocco / Western Sahara	5,700 ,000	50,000,000
China	3,700,000	3,300,000
Algeria	-	2,200,000
Syria	100,000	1,800,000
Brazil	260,000	1,700,000
South Africa	1,500,000	1,500,000
Saudi Arabia	-	1,400,000
Egypt	100,000	1,300,000
Jordan	1,500,000	1,300,000
Australia	82,000	1,100,000
United States	1,100,000	1,000,000
Russia	200,000	600
Global reserves	**15,627,000**	**69,904,700**

Despite the efforts of the USGS to sustain the share of information about global phosphate rock reserves and their distribution, there is limited information about the reserves in some countries like China and Tunisia [215]. Therefore, the information for some countries in the USGS database is based on estimations [213].

This is the case of Tunisian phosphate rock reserves. The reserves size reported by the USGS is 100 MMT since the last update in 1999. In the next section, the phosphate reserves in Tunisia will be updated according to the data collected in 2018 and provided by the Company of Phosphate of Gafsa (CPG).

3.3.2 The update of Tunisian phosphate rock reserves

The following reporting of the Tunisian reserves is based solely on the data provided by the Company of Phosphate of Gafsa "Compagnie des Phosphate de Gafsa" (CPG). It is a state-owned company, and it is responsible for research and development in the sector of phosphate rock exploration, extraction, and mining. In respect to the secrecy requested by the company, the mining sites' locations and names were not shared. The active mining sites are designated as 'Sector', and the current projects with measured reserves are indicated as 'Project'.

In 2011, new phosphate deposits in Tunisia were declared as indicated reserves in the mining area of Gafsa, in the region of Tozeur, and the region of Sra Ourtane, located in the northwest of the country. These deposits lead to a sharp increase in the reserves of the country. The phosphate deposit of Tozeur has been for long considered as non-profitable material as the identified deposit has a low-grade concentration of 12% P_2O_5. However, the reverse flotation technique would increase the content of beneficiated phosphate rock to 27 % P_2O_5, according to the joint study of CPG and the Faculty of Sciences of Tunis [17].

According to the geological department at the (CPG), the current demonstrated resources are more than 6 billion metric tons. The reserves base (indicated as measured reserves in the Annex C, Table C-2) was evaluated to be 3.3 billion metric tons in 2017 compared to 100 MMT reported by USGS [212]. Figure D-1 (a) and (b) show an active and dynamic exploration of new reserves in the current mining sites. The measured resources are declining in all active sectors except for sector 3 (from 59 MMT in 2011 to 104 MMT in 2018). The demonstrated resources in the current mining sites vary between 100-290 MMT per sector, with a total of 1 billion metric tons of demonstrated reserves in the current active mining area.

In addition to the potential of the current active mining area, the new projects bring more than 5 billion metric tons of demonstrated reserves (see Figure 3.2 (c) and (d)), including 2.8 billion metric tons of measured reserves. The most significant size of the measured reserves is project number 3. Project number 3 is located in the area of Sra Ourtane. It was highly mediatized as it promises to increase the mining capacity of the company, which declined after 2010 due to the regional socio-political instability.

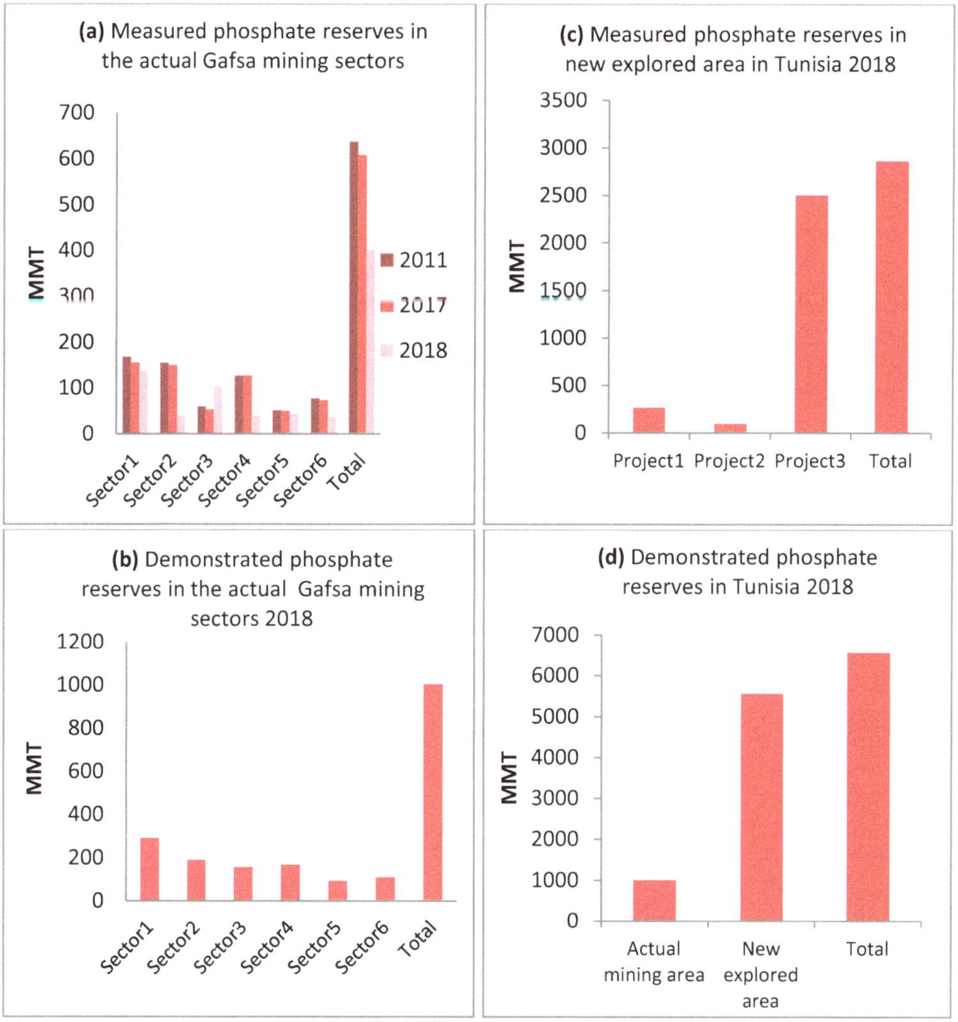

Figure 3.2: **Evolution of demonstrated resources and measured reserves between 2011 and 2018 in Tunisia; based on data from the geological department at the CPG.**

3.3.3 The update of global phosphate reserves and the evolution of the depletion time

The Tunisian reserves update leads to an update of the global reserves from 69.9 billion metric tons, according to the USGS database in 2017, to 73.1 billion metric tons for the same year. Since 2010, the global reserves of phosphate were not alarming anymore. However, Figure 3.2 shows that the growth rate of the global demand for the production of beneficiated phosphate rock remains faster than the growth rate of global reserves. The Average Annual Growth Rate (AAGR) calculated for 2010-2018 is 3% for the reserves and 6% for the global demand. Thus, the global demand rate is two times higher than the reserves update rate.

Moreover, Figure 3.4 displays the depletion time evolution, which is the ratio of reserves to consumption (R/C) (see definitions section 3.2.3). The depletion time is expressed in years. Since 2011, the depletion time is decreasing very slightly with a slop of (-0.07) degree. This suggests that despite the constant increase of the global demand, the reserves update is dynamic to keep the depletion time more or less steady between 250-300 years.

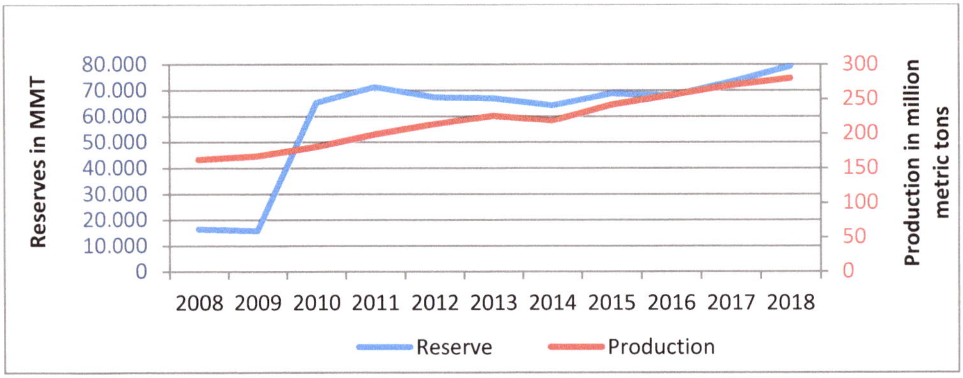

Figure 3.3: Global growth rate of phosphate rock reserves and global demand expressed by the mine production phosphate rock in 10 years based on data from USGS [203–213] and CPG (2017-2018)

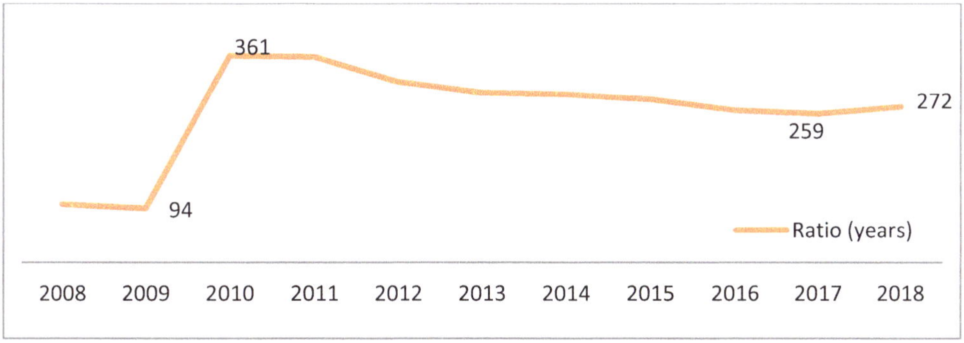

Figure 3.4: Reserves to Production ratio evolution in 10 years based on the data from USGS [203–213] and CPG for Tunisia (2017-2018) [136]

3.3.4 The dynamic evolution of global phosphate reserves

The current global phosphate rock reserves would last 270 years with the current consumption rate. Nonetheless, as it is demonstrated in the previous section, the depletion time fluctuates due to the dynamic updates of the reserves. Figure 3.5 shows the availability of phosphate rock reserves in each country through a period of ten years. Before 2010, the global reserves were 15.6 billion metric tons. The update of the Moroccan[1] reserves from 5.7 billion metric tons in 2009 to 50 billion metric tons changed the fact that phosphate rock was a scarce resource. The depletion time was estimated at 50-100 years in 2009 [39]. In Figure 3.5, the black line marks the update of global reserves in 2010. It represents the end of the theory of Peak of Phosphorus in 2035.

Moreover, other resource discoveries are occurring worldwide, like in the case of Tunisia, Egypt, Saudi Arabia, and Brazil, which can compensate for the decline of reserves in some parts of the world, such as Iraq, the USA, and Russia (seeFigure 3.5). The resources and reserves of phosphate are not static. According to Scholz et al. (2013), due to the enabling factors such as technology and economic environment, new resources can be discovered, and identified resources can be transformed into reserves [175].

[1] In this work the terms Moroccan reserves and Morocco as it is marked in figures represent Morocco and Western Sahara reserves so that the text in the figures is less cumbersome.

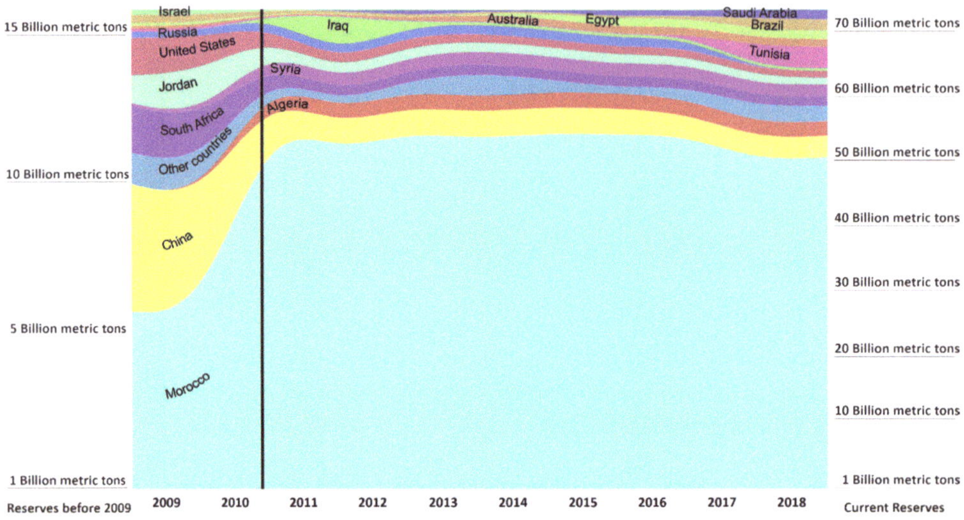

Figure 3.5: The dynamic evolution of global phosphate rock reserves for ten years, based on the data of USGS [203–213] and the CPG for Tunisia (2017-2018) [136], the graph representation of reserves is relative to the size of the country's reserves

The USGS has reported that the global resources of phosphate rock are more than 300 billion metric tons [213]. The current reserves represent 24% of the global resources. The current depletion time of the resources according to the current consumption rate is 1,113 years. However, if we suppose the following scenario: resources are static (which means no new identified resources), the reserves annual growth rate is 3% (see Figure 3.6-a), and the production continues to grow 6% per year (see Figure 3.6-b), then the 300 billion metric tons identified resources would turn into reserves by the year 2063. The depletion time would be 77 years in 2063 (see Figure 3.6-c). Further information is presented in Annex C, Table C-2. Thus, the resource dynamic is good news to reduce the risk of phosphate rock scarcity but other factors, such as the mineral concentration of the ore and the global demand growth rate, which would put heavy pressure on phosphate rock resource availability. The need for more minerals means the recovery of low-grade ores with high heavy metals content and the allocation of more water and energy resources for their extraction. Therefore, the increase in the global demand for phosphorus will create more technical and environmental challenges.

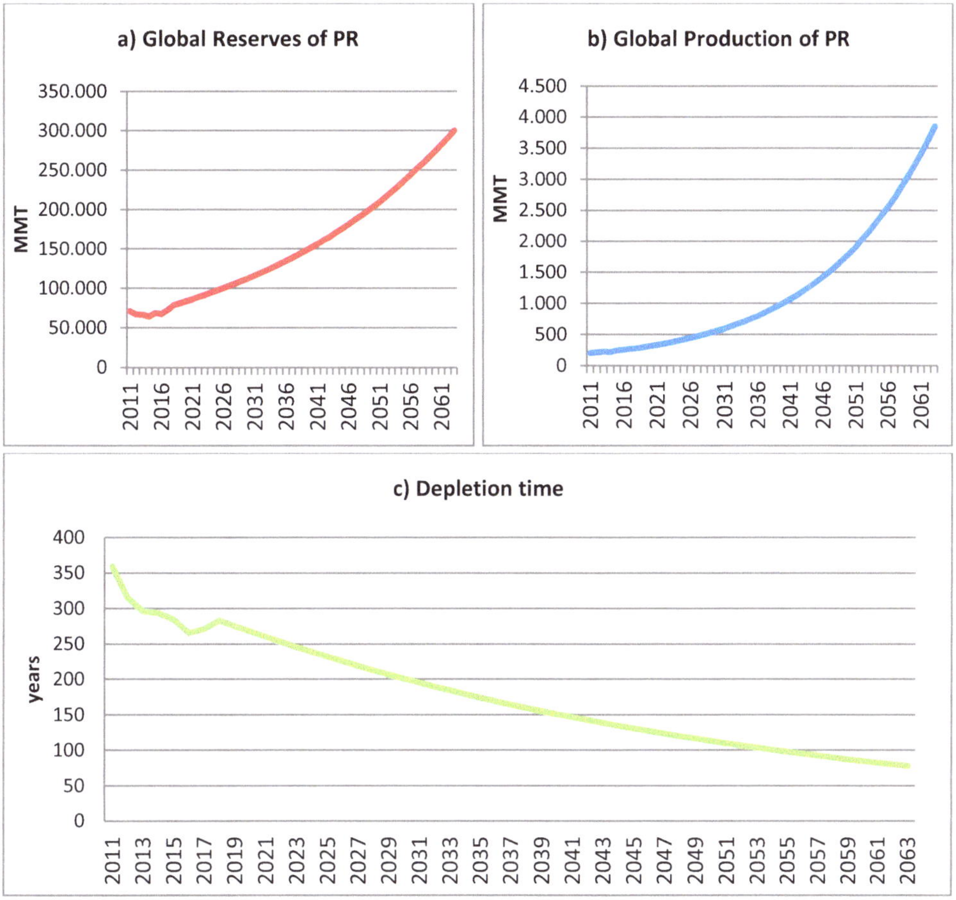

Figure 3.6: **Global reserves (a), global production (b), and the depletion time forecast (c) in case of static resources scenario and 3% AAGR of the reserves and 6% AAGR of the production**

3.3.5 Geo-concentration of the global reserves

Phosphate rock reserves exist in every region of the world. However, the most abundant sedimentary deposits are found in North Africa (55.5 billion metric tons), and the most extensive igneous phosphate occurrences are located in Brazil (1.7 billion metric tons), Finland (1 billion metric tons), South Africa (1.5 billion metric tons), Canada (76 million metric tons) and Russia (600 thousand metric tons) [212].

The sedimentary phosphate reserves are more significant than the igneous phosphate reserves.

Figure 3.7 shows that after updating the Tunisian reserves, the weight of North Africa's share of the global reserves increased from 74.8% to 76%. Morocco has the largest global reserves (50 billion metric tons), representing 68% of global reserves, followed by Tunisia (3.32 billion metric tons).

The second most significant deposits are in the Middle East, with a total of 6.23 billion metric tons of phosphate rock reserves, which represent 9% of the global reserves. Syria has the largest share in the Middle East (1.8 billion metric tons), followed by Saudi Arabia (1.4 billion metric tons), which has seen a steady increase in its reserves and production capacity since 2012.

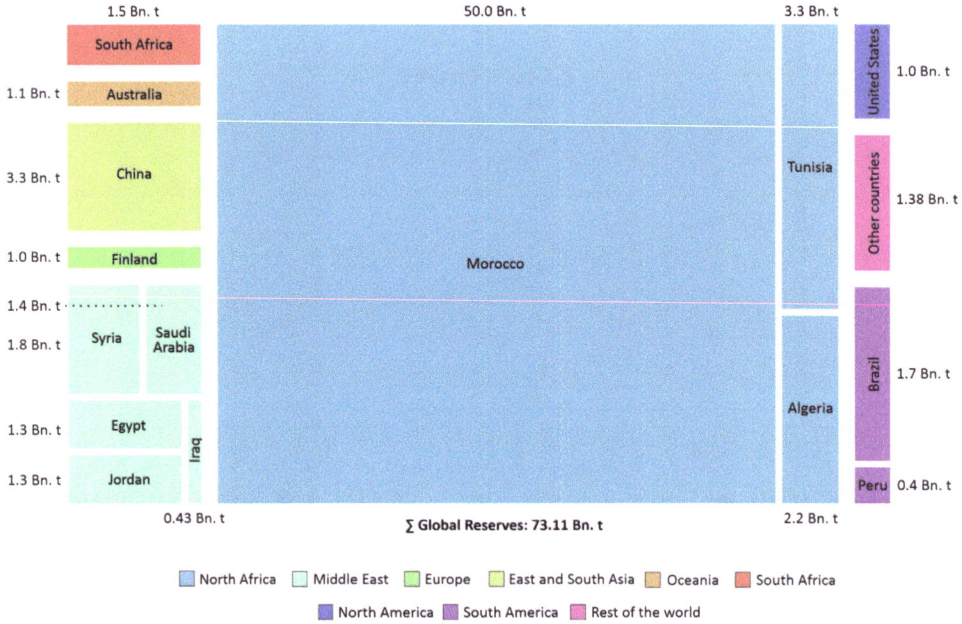

Figure 3.7: Geo-concentration of the global phosphate rock reserves (2017) expressed in billion metric tons (Bn. t) based on the data of USGS [203–213] and CPG for 2017 [136].

In the East and South Asia region, 5% of global reserves are located mainly in China, which has the third-largest global reserves (3.3 billion metric tons). The rest of the global reserves are distributed more or less equally between the region of South America (3%), sub-Saharan African countries (2%), North America (2%), Australia (2%), and Europe (1%). The Australian reserves reported by the USGS in 2017 were estimated to be 1.1 billion metric tons [212]. However, the official data published by Geoscience Australia reported

289 MMT of reserves and 1.072 billion metric tons of economic demonstrated resources in 2016 [20].

3.4 The global flow of phosphorus

3.4.1 The global production of beneficiated phosphate rock

The global production of beneficiated phosphate rock in 2017 was 269.5 MMT. It continues to grow by the average annual growth of 6% calculated from 2010 to 2018 (see Figure 3.3). Figure 3.8 shows the leading countries producing beneficiated phosphate rock. The top 3 producers in 2017 are China (144 MMT), Morocco (30 MMT), and the USA (27 MMT). The Top 3 producers represent 75% of the global production. The USGS reported that the industry analysts estimated the Chinese production of beneficiated phosphate rock to be around 85 MMT, considerably lower than the official data published by the Government of China [212].

Nonetheless, China is the world's largest producer of beneficiated phosphate rock. Smaller producers with an annual production of less than 10 MMT are Jordan, Brazil, Saudi Arabia, Tunisia, Egypt, and Israel. Like the reserves, the production of beneficiated phosphate rock is highly concentrated in the same regions of North Africa, the Middle East, China, and North and South America.

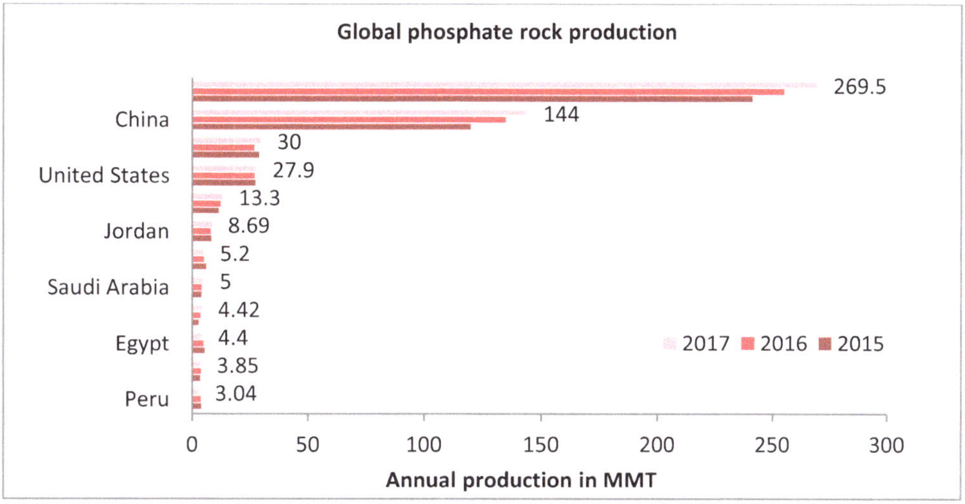

Figure 3.8: Global phosphate rock production 2015-2017 based on the data of USGS [210–212] and CPG production reports (2015-2017) [53, 55]

3.4.2 The international trade of beneficiated phosphate rock

According to the data of the International Trade Center (ITC), the size of the global market of beneficiated phosphate rock was 25.3 MMT in 2016 and 27.7 MMT in 2017 [116]. The global market represents 10% of the global production of beneficiated phosphate rock. Moreover, according to Figure 3.8, world production increased from 2015 to 2017, while the quantity of beneficiated phosphate rock exports on the global market was relatively constant.

This is explained by the fact that the domestic consumption of the primary producers is very high. For instance, China, the world's largest producer of beneficiated phosphate rock, has a share of less than 1% of the total global exportation. The United States, the world's third-largest producer are net importers. They need more phosphate rock than they produce to meet their domestic needs (see Figure 3.9).

Besides, Figure 3.9 exhibits international trade between countries in 2016. The top 10 exporting countries represent 64% of the global export market. The leading exporters on the global market are Morocco, with 19% of the global exports, followed by Jordan (16%) and Peru (13.5%). Morocco is mainly exporting to India, Brazil, Europe (Poland), Turkey, Mexico, and New Zealand. Jordan is mainly exporting to India and Indonesia, while Peru is mainly exporting to the United States, Brazil, and India.

The weight of the top 10 importing countries is 75% of the global import market. The leading importers are India (7.5 MMT), which represents a third of the global market, Brazil (1.71 MMT), Indonesia (1.69 MMT), and the United States (1.6 MMT). The European Union imported approximately 5 MMT in 2016 (see Annex C, Table C-3). As an economic union, the EU is the second-largest importer of beneficiated phosphate rock after India (see Figure 3.9). The European countries belonging to the top 10 importing countries are Poland, Lithuania, and Belgium, which together imported 12% of the global imports and 69% of total European imports (in Figure 3.9, Poland, Lithuania, and Belgium are represented together as Europe). The main exporters to the European Union in terms of quantity are Morocco, which is mainly exporting to Poland, and Russia, which is mainly exporting to Belgium and Lithuania.

The importing countries are those which do not have sufficient phosphate rock reserves to meet their domestic markets (such as India), which do not have phosphate rock deposits at all (such as the European countries except Finland), and countries which have a big fertilizer industry and an agricultural need for fertilizers (such as the United States, Brazil, and Indonesia). In terms of consumption, China consumes all of its production, which is three to five times the size of the global market (depending on the consideration of either 85 MMT, as estimated by industry analysts, or 144 MMT, as the official data published by the Chinese government [212]). China uses its production for downstream industries such as phosphoric acid production, white phosphorus production, and fertilizer production.

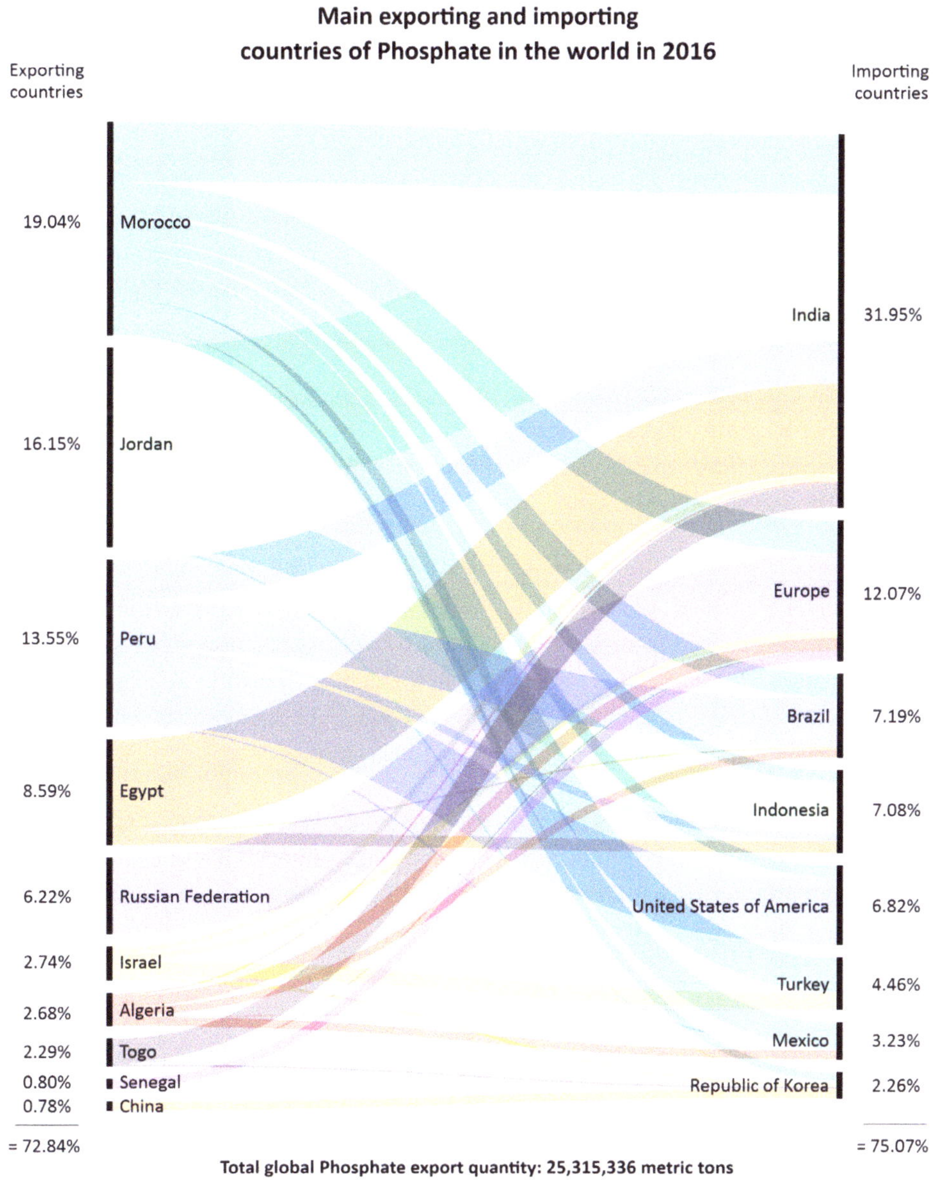

Figure 3.9: **The global trade of beneficiated phosphate rock based on the data from the International Trade Center (ITC) for 2016**

3.5 Conclusion

After the update of the Tunisian phosphate rock reserves, the total global phosphate rock reserves add up to 73.1 billion metric tons in 2017 compared to 69.9 billion metric tons, reported by the USGS.

The phosphate rock as a primary source of phosphorus is highly geo-concentrated in three regions North Africa, the Middle East, and China. The world's top 3 countries- having the most extensive reserves- are Morocco with 50 billion metric tons, Tunisia with 3.32 billion metric tons, and China with 3.3 billion metric tons. Phosphate rock resources and reserves are sufficient to meet the current consumption rate. The current depletion time is 272 years, and it is more or less stable, evolving from 360 to 272 years between 2010 and 2018. However, the global demand continues to grow at a rate twice as high as the reserves growth rate.

The depletion time is an early indicator to assess the availability of a mineral. Nevertheless, it is not an accurate indicator of resource scarcity as the phosphate rock consumption is demand-driven. Efforts to improve the efficiency of phosphate use through the value chain would delay the depletion time.

As the reserves of minerals are dynamic, they can increase due to enabling factors such as economic, technological, environmental, social, and legal factors [175]. The geological discovery of new resources can also contribute to converting resources into recoverable reserves. However, the content of phosphorus in the deposit is highly variable and tends to decrease. Therefore, the available stock of minerals would decrease, as the ore grade quality is declining, and the demand growth rate is increasing.

This leads to challenges such as (1) an increasing amount of heavy metal, specifically the cadmium in beneficiated phosphate rock, (2) a massive amount of water and energy usage for phosphate rock beneficiation, and (3) the losses of phosphorus causing eutrophication through the value chain.

4 Global supply of phosphorus and criticality analysis

4.1 Introduction

The geological scarcity of phosphate rock was disproved in Chapter 3. Nevertheless, the criticality of phosphorus is beyond the assessment of geological factors. Criticality assessment also includes other factors, which could increase the supply risk of economically important material.

In this chapter, phosphorus criticality will be investigated by screening the phosphorus supply bottlenecks along the global supply chain, followed by identifying the factors that affect phosphorus criticality on the global supply chain level.

4.2 The global flow of phosphorus along its supply chain

4.2.1 Data sources

The flow of phosphorus will be mapped along the global supply chain like it was suggested by Gantner (2016) [82]. The key-intermediate phosphate products in the value chain are: (1) the beneficiated phosphate rock, (2) the phosphoric acid (H_3PO_4), an intermediate in fertilizer production, (3) the main fertilizers (DAP, MAP, and TSP), and (4) the white phosphorus (P_4). Based on these three phosphate products, all derivatives of Phosphorus are produced. The choice for these key-intermediate phosphate products is according to the size of the global production and consumption and their crucial role in the value chain.

To standardize the flow of phosphorus from different key-intermediate phosphate products, the data provided in this part is on a nutrient base. Thus, the supply chain of phosphorus is expressed in Phosphorus Pentoxide (P_2O_5).

For the beneficiated phosphate rock, the conversion factor to P_2O_5 is 0.29 (the average content of beneficiated phosphate rock is 29%). The conversion factor[1] from phosphoric acid (H_3PO_4) to P_2O_5 is 0.54 (the marketable product of one metric ton of phosphoric acid containing 75% of H_3PO_4). For the fertilizer Diammonium Phosphate (DAP), the conversion factor from the marketable product to the nutrient P_2O_5 is 0.46. For the fertilizer Monoammonium Phosphate (MAP), the conversion factor from the marketable product to the nutrient P_2O_5 is 0.52. The conversion factor[2] from the marketable product Triple Superphosphate (TSP) to the nutrient P_2O_5 is 0.46. The conversion factor[3] from marketable product white phosphorus (P_4) (at a purity of 98% -99%) to the nutrient P_2O_5 is 2.2.

The data used to calculate the production of the beneficiated phosphate rock are from the USGS database [211, 212]. The data used to calculate the production of the phosphoric acid, the DAP, MAP, and TSP are from the Food and Agriculture Organization (FAO) [77, 79] and the International Fertilizer Association (IFA) [111]. Furthermore, the data used for the global trade of different phosphate derivates are from the International Trade Center (ITC) [116] and the IFA [112].

The global production of white phosphorus (P_4) was estimated according to the critical raw material factsheet report prepared for the European Commission [19]. The global production between 2010 and 2014 was declared 950 thousand metric tons [19]. According to the Market analysis report of IHS Markit (2017), China produces 87% of global production [101]. It increased its production capacity between 2002 and 2012 by 9%. Meanwhile, a sharp decline in the production capacity in Europe and North America took place. Therefore, the global white phosphorus production is estimated to be 1 MMT in 2016 and 2017.

The data used for the flow of white phosphorus (P_4) and its derivate products Phosphorus Trichloride (PCL_3) and the other chemicals are from the Market analysis report of IHS Markit (2017) [101].

---

[1] The molar mass of P_2O_5 is 142 g, the molar mass of H_3PO_4 is 97.995g; $1P_2O_5+3H_2O \rightarrow 2H_3PO_4$; $1P_2O_5$*142/$2H_3PO_4$*97.995, the conversion factor i= (142/196)*0.75(concentration of H_3PO_4 in the marketable product)=0.54

[2] The conversion factor for phosphoric acid, DAP, MAP and TSP are taken from IAF fertilizer converter page: https://www.ifastat.org/converter/fertilizer-converter/

[3] The molar mass of P_2O_5 is 142 g and the molar mass of P_4 is 124 g; $1P_4+5O_2 \rightarrow 2P_2O_5$; $2P_2O_5$*142/$1P_4$*124; the conversion factor = 284/124 = 2.29

4.2.2 Data analysis

Figure 4.1 represents the phosphorus flow (expressed in terms of nutrient P_2O_5) between the key-intermediate phosphates products along the global supply chain. The flow percentage is represented in purple circles. It shows the flow amount and direction between key-intermediate phosphates products (the blue boxes). The beneficiated phosphate rock is used predominantly to produce phosphoric acid, which is the intermediate product for fertilizers production. According to Figure 4.1, 65% of phosphorus is used to produce phosphoric acid, and 88% of the produced phosphoric acid is used to produce fertilizers globally.

Despite the increase of the beneficiated phosphate production, the exports on the global market are very small, and the massive consumption of beneficiated **phosphate rock** occurs in domestic markets of producing countries, especially in the largest phosphoric acid producing countries such as China, the United States, and Morocco (see Table 4.1).

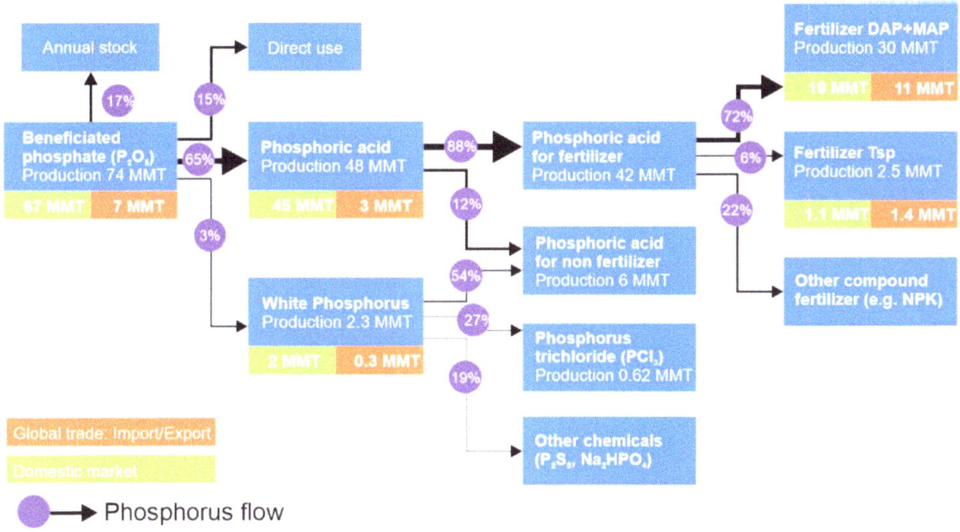

Figure 4.1: Global flow of phosphorus from primary resources along the global value chain expressed in million metric tons (MMT) of P_2O_5 for the year 2017

According to Figure 4.1, the main producers of beneficiated phosphate rock prioritize their domestic markets, except for fertilizers (this accounts for around 50% of the global production). The countries rich in phosphate rock invest in downstream industries. Especially China is targeting more down-streaming of the phosphorus value chain; they are the world's leader in beneficiated phosphate production, phosphoric acid production, phosphate fertilizer production, and white phosphorus production.

Furthermore, the production of the purified phosphoric acid, usually produced in an electrothermal process using white phosphorus as input, is declining. Its production from the wet process became more competitive than the electrothermal process. Figure 4.1 shows that the flow of phosphorus from the wet process is 12% (marked in the figure as phosphoric acid for non-fertilizer uses), and 54% from the electrothermal process. According to IHS Markit (2017), the global market of thermal phosphoric acid is declining by 1.5% per year [101]. This technology change will increase the geo-concentration of phosphate products in upstream of the global supply chain, and the purified phosphoric acid for non-fertilizer use would be more concentrated in countries increasing their phosphoric acid production capacity.

Table 4.1: Global production and trade of beneficiated phosphate rock, phosphoric acid, and main fertilizers expressed as P_2O_5 in 2016 and 2017. Data source: [77, 79] [111, 112] [116] [211, 212]

	Production			Export			Import		
	2016	**2017**	**Top 3 producers**	**2016**	**2017**	**Top 3 Exporters**	**2016**	**2017**	**Top 3 importers**
Beneficiated phosphate rock (MMT P_2O_5)	74	78	China/ Morocco/ USA	7.3	8	Morocco/ Jordan/ Peru	7.0	7.0	India/ Brazil/ Indonesia
Phosphoric acid (MMT P_2O_5)	48.4	49.6	China/ USA/ Morocco	3.2	3	Morocco/ USA/ Tunisia	3.3	3.0	India/ Nether-land/ Turkey
DAP+MAP (MMT P_2O_5)	29.6	31.0	China/ India/ USA	11.4	11.9	China/ Morocco/ USA	11.6	12.6	India/ Pakistan/ Vietnam
TSP (MMT P_2O_5)	2.5	2.6	Egypt/ Tunisia/ China	1.4	1.8	Tunisia/ Mexico/ Bulgaria	1.4	1.8	Brazil/ Cuba/ Ukraine
White phosphorus (MMT P_2O_5)	2.30	2.30	China/ USA/ Kazakhstan	0.37	0.5	Vietnam/ Kazakhstan/USA	0.3	0.3	Germany/ India/ USA/Japan

4.3 Supply risk assessment of phosphorus along the global value chain

The risk of supply restriction of the key-intermediate phosphate products (1) Beneficiated phosphate rock, (2) Phosphoric acid, and (3) White phosphorus were estimated according to a calculation formula based on the Geo-concentration of the production, the socio-political stability of the country, the resources governance of the country and the environmental performance of the country. The parameters, their indicators, and the references are shown in Table 2.2 in section 2.3.1.

4.3.1 Calculation formula

For each phosphate product along the value chain, the supply risk related to socio-political stability ($S.R_{Socio\text{-}Political}$) was calculated by multiplying the country concentration (HHI) with the political stability of the country (WGI_{PSAV}). This approach was adopted from the European Commission (EC) [69, 72].

The Supply risk related to the resource governance ability of a country ($S.R_{governance}$) was calculated by multiplying the country concentration (HHI) with the Resource governance index (RGI). The supply risk related to environmental performance ($S.R_{Environmental}$) was calculated by multiplying the country concentration (HHI) with the environmental performance index (EPI).

In order to align the scale direction of the WGI_{PSAV} to the HHI scale, the scale of the WGI_{PSAV} (-2.5 - 2.5) was shifted to positive values and normalized to a decimal scale (0.01 - 1.00) according to formula **(1)**. The scale direction of RGI and EPI were also aligned to the HHI scale by inverting the scales from (0-100) to (0.01 - 1.00), according to the formulas **(2)** and **(3)**.

$$\textbf{(1)} \quad S.R_{socio-political} = \sum_{i=1}^{n}(country's\ global\ share/100)_i^{\ 2} * \left(\left(-WGI_{(PSAV)i} + 2.5\right) * 0.2\right)$$

$$\textbf{(2)} \quad S.R_{governance} = \sum_{i=1}^{n}(country's\ global\ share/100)_i^{\ 2} * \left(1 - \frac{RGI_i}{100}\right)$$

(3) $$S.R_{Environmental} = \sum_{i=1}^{n} (country's\ global\ share/100)_i^{\ 2} * \left(1 - \frac{EPI_i}{100}\right)$$

- S.R = the supply risk
- WGI_{PSAV} = the World Governance Indicator on the political stability and absence of violence of a country
- RGI = the Resource governance Index measuring the ability of a country to manage resources
- i= the number of countries sharing 95% of global production or global export

4.3.2 Results

4.3.2.1 The supply risk of beneficiated phosphate rock

- **Global production**

The results in Table 4.2 show that the beneficiated phosphate rock production is very highly concentrated (HHI>0.18). Fourteen countries share around 96% of global production (see), which is concentrated mainly in China (50%), Morocco (11%), and the USA (10%).

Moreover, both the supply risk related to the socio-political stability of the countries (WGI_{PSAV}) and the environmental performance of the countries (EPI) are at a high-risk level. The supply risk related to resource governance (RGI) is at a medium-risk level.

Table 4.2: Beneficiated phosphate rock geo-concentration of the production and global supply risk analysis (2017)

	HHI	$S.R_{socio\text{-}political}$	$S.R_{Governance}$	$S.R_{Environmental}$
Production of beneficiated phosphate rock	0.31	0.17	0.14	0.15
	Very highly concentrated	High supply risk	Medium supply risk	High supply risk
Very low risk: 0.0<x<0.05 / **Low risk:** 0.05≤x<0.10 / **Medium risk:** 0.10≤x<0.15 / **High risk:** 0.15≤x<0.18 / **Very high risk:** ≥0.18				

Table 4.3: Beneficiated phosphate rock global production share per country and their supply risk indicators in 2017

Country	Global share	WGI_{PSAV}	RGI	EPI
China	53.4%	-0.25	55[a]	50.74
Morocco	11.1%	-0.41	37	63.47
United States	10.4%	0.63	74[a]	71.19
Russia	4.9%	-0.67	45[a]	63.79
Jordan	3.2%	-0.53	40[b]	62.2
Brazil	1.9%	-0.41	71[a]	60.7
Saudi Arabia	1.9%	-0.62	36[a]	57.47
Egypt	1.6%	-1.42	39[a]	61.21
Tunisia	1.6%	-1.05	56[a]	62.35
Israel	1.4%	-0.88	60[b]	75.01
Peru	1.1%	-0.26	62	61.92
Australia	1%	0.9	71	74.12
Vietnam	1%	0.31	48[a]	46.96
South Africa	1%	-0.27	57	44.73
Global share	95.7%	n/a	n/a	n/a

a): The resources governance index was calculated for the Oil gas extractive sector

b): For Jordan and Israel, there are no data in the NRGI database. For Jordan, the value was estimated comparing to countries in the same region with a comparative level of WGI, which are Egypt and Saudi Arabia. For Israel, the value was estimated comparing to countries with a comparative level of WGI, which is South Africa

- **Global exportation**

The results in and Table 4.5 show that the global market of beneficiated phosphate rock is also very highly concentrated (HHI> 0.18) but less concentrated than the global production of beneficiated phosphate rock. Ten countries share 96% of global exports (see Table 4.4). Morocco shares 40% of the global export market, followed by Jordan (18%) and Peru (11%).

The supply risk related to socio-political stability (WGI_{PSAV}) and the supply risk related to resource governance capability (RGI) are both at a medium-risk level. The supply risk related to environmental performance (EPI) is at a low-risk level. The difference in the supply risk level between the global production and the global export market is explained by the weight of China as the first producer of beneficiated phosphate rock (53.4% of the global production) but has a minimal share of the global export market. China has a weak WGI_{PSAV} (-0.25) and a weak EPI (50.74). However, Morocco is the largest exporter of beneficiated phosphate rock on the global market has a weak WGI_{PSAV} (-0.41) but a significantly better EPI (63.47).

Table 4.4: Beneficiated phosphate rock global exports market share per country and their supply risk indicators in 2017

Exporters	Global share	WGI_{PSAV}	RGI	EPI
Morocco	40%	-0.41	37	63.47
Jordan	18%	-0.53	40[b]	62.2
Peru	11%	-0.26	62	61.92
Russian	9%	-0.67	45[a]	63.79
Egypt	4%	-1.42	39[a]	61.21
Algeria	4%	-0.96	33[b]	57.18
Togo	3%	-0.74	54[b]	41.78
Israel	2%	-0.88	60[b]	75.01
Senegal	2%	-0.04	40[b]	49.52
Kazakhstan	2%	0.02	56	54.56
Global export	96%			

a): The resource governance was calculated for the Oil gas extractive sector

b): For Jordan. Israel. Togo and Senegal, there is no data in the NRGI database; for Jordan, the value was estimated comparing to countries in the same region with a comparative level of WGI, which are Egypt and Saudi Arabia. For Israel, the value was estimated comparing to countries with a comparative level of WGI, which is South Africa. For Togo, the value was estimated comparing to countries in the same region with a comparative level of WGI, which are Burkina Faso and Ghana. For Senegal, the value was estimated comparing to countries in the same region with a comparative level of WGI, which are Sierra Leone and Guinea [144, 223]

Table 4.5: Beneficiated phosphate rock global export market concentration and supply risk analysis (2017)

	HHI	$S.R_{socio\text{-}political}$	$S.R_{Governance}$	$S.R_{Environmental}$
Exportation of beneficiated phosphate rock	0.22	0.13	0.13	0.08
	Very highly concentrated	Medium supply risk	Medium supply risk	low supply risk
Very low risk:0.0<x<0.05 / **Low risk**: 0.05≤x<0.10 / **Medium risk**: 0.10≤x<0.15 / **High risk:** 0.15≤x<0.18 / **Very high risk:** ≥0.18				

4.3.2.2 The supply risk of phosphoric acid

Concerning the export of phosphoric acid on a global level, Table 4.6 shows that the global market is moderately concentrated (HHI<0.18). Twelve countries share 95% of the global market (see Table 4.7).

The supply risk related to the socio-political stability of the country (WGI_{PSAV}) is at a medium-risk level. The supply risk related to the environmental performance of the country (EPI) is at a low-risk level. This is explained by the fact that the export of phosphoric acid on the global market is less concentrated than the export of beneficiated phosphate rock; the HHI is 0.17 compared to the export of beneficiated phosphate rock, which has an HHI of 0.22. However, phosphoric acid is mainly exported by countries controlling reserves and the production of beneficiated phosphate rock.

Table 4.6: Phosphoric acid global export market concentration and global supply risk (2017)

	HHI	$S.R_{socio\text{-}political}$	$S.R_{Environmental}$
Exportation of Phosphoric acid	0.17	0.10	0.06
	Moderately concentrated	Low supply risk	Low supply risk
Very low risk: 0.0<x<0.05 / **Low risk:** 0.05≤x<0.10 / **Medium risk:** 0.10≤x<0.15 / **High risk:** 0.15≤x<0.18 / **Very high risk**: ≥0.18			

Table 4.7: Phosphoric acid global export market shares per country and their supply risk indicators in 2017

Exporter	Global share	WGI_{PSAV}	EPI
Morocco	35%	-0.41	63.47
United States of America	12%	0.63	71.19
Tunisia	11%	-1.05	62.35
Israel	8%	-0.88	75.01
Senegal	6%	-0.04	49.52
South Africa	6%	-0.27	44.73
Jordan	5%	-0.53	62.2
Belgium	5%	0.42	77.38
Netherlands	3%	0.92	78.64
Vietnam	2%	0.31	46.96
Finland	1%	1.07	78.64
Mexico	1%	-0.65	59.69
Global share	95%		

4.3.2.3 The supply risk of white phosphorus

For the global market of white phosphorus, Table 4.8 shows that the global market is very highly concentrated (HHI>0.18). Two main countries (Vietnam and Kazakhstan) share almost 80% of the exports on the global market (see Table 4.9). Thus, the supply risk related to the socio-political instability (WGI_{PSAV}) and the environmental performance of the country (EPI) are both at high risk level.

Table 4.8: White phosphorus global export market concentration and supply risk analysis (2017)

	HHI	$S.R_{socio\text{-}political}$	$S.R_{Environmental}$
Exportation of white phosphorus	0.32	0.15	0.15
	Very high concentrated	High risk	High risk
Very low risk:0.0<x<0.05/Low risk: 0.05≤x<0.10/Medium risk: 0.10≤x<0.15/High risk: 0.15≤x<0.18/Very high risk: ≥0.18			

Table 4.9: White phosphorus global export market shares per country and their supply risk indicators in 2017

Exporters	Global share	WGI_{PSAV}	EPI
Vietnam	42%	0.31	46.96
Kazakhstan	36%	0.02	54.56
United States of America	8%	0.63	71.19
Poland	5%	0.52	64.11
China	3%	-0.25	50.74
Germany	3%	0.58	76.37
Belgium	2%	0.42	77.38
Netherlands	1%	0.92	78.38
Global share	99.7%		

4.3.2.4 Interpretation

Figure 4.2 shows the aggregation of the supply risk factors into one single value. The highest supply risk concerns the global production (S.R=0.46) and export of beneficiated phosphate rock (S.R=0.34), mainly due to weak natural resources governance (RGI) and high risk of socio-political instability (WGI_{PSAV}) in the producing countries. Also, the global production of beneficiated phosphate rock is very highly concentrated in countries with low environmental performance (EPI).

In comparison to the phosphoric acid exports (S.R=016), white phosphorus has a higher supply risk (S.R=0.30) related to environmental performance ($S.R_{EPI}$ = 0.15) as well as socio-political instability (S.RWGI = 0.15). This is explained by the higher geo-concentration of the production in two countries, Vietnam and Kazakhstan.

The supply chain of phosphorus is highly concentrated for all the key-intermediate phosphate products (beneficiated phosphate rock, phosphoric acid, and white phosphorus), and the supply risk is located on both the upstream and the downstream of the value chain. The highest risk concern the production of the primary source of phosphorus (beneficiated phosphate rock). The main risk factors are the socio-political stability of the countries and their resource governance, including environmental performance.

Furthermore, in Figure 4.2, it can be noticed that the risk related to socio-political instability and the environmental performance of countries evolve in the same direction; the higher the socio-political risk, the lower the environmental performance. This suggests the existence of a certain degree of correlation between both risk factors.

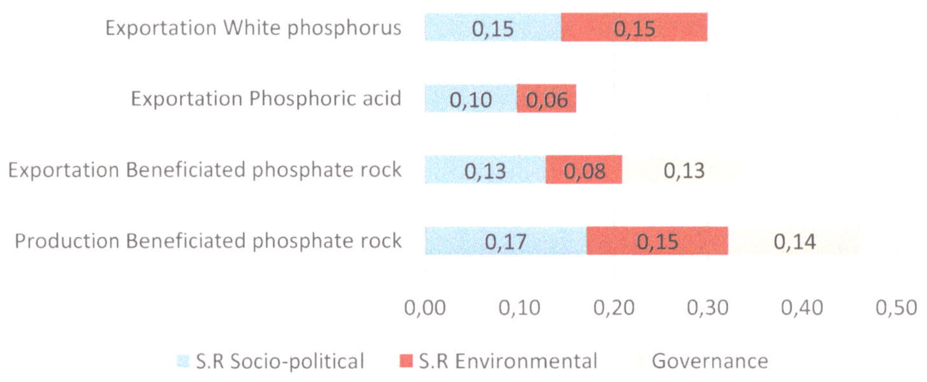

Figure 4.2: Comparison of the global supply risk level of key intermediate phosphate products; (1) Beneficiated phosphate rock, (2) Phosphoric acid, and (3) White phosphorus

4.4 Conclusion

To conclude this chapter, first, the global supply chain of phosphorus is very concentrated in the upstream part due to the concentration of the reserves and the downstream part as two countries share 80% of the global export market of white phosphorus. The supply risk analysis shows that the regions that control the supply chain of beneficiated phosphate rock and phosphoric acid are the same regions that control the reserves (North Africa, the Middle East, and China).

Second, the supply risks analysis results are similar to the European commission critical raw material studies [69, 70, 73] as both beneficiated phosphate rock and white phosphorus are subject to a high risk of supply interruption.

Third, the factors that can increase the criticality of phosphate rock are the socio-political situation, the governance of natural resources concerning the beneficiated phosphate rock, and the producing countries' environmental performance.

A very relevant and interesting observation is the socio-political and environmental supply risk factors tend to evolve in the same direction, which suggests a link between the socio-political instability and low environmental performance of the country.

5 Framework conditions of phosphate rock mining in Gafsa, Tunisia

5.1 Introduction

Tunisia is one of the world's top-ten producers of beneficiated phosphate rock and home to the world's fifth-largest phosphate rock producing company, the state-owned "Compagnie de Phosphate de Gafsa" (CPG). Tunisia's mining sector scores 46 out of 100 points in the 2017 Resource Governance Index, placing the country 48th among the list of 89 assessed countries in the overall ranking. Despite the overall weak performance, Tunisia's mining sector ranks fifth in the Middle East and North Africa (MENA) region. The phosphate mining industry is prone to environmental and social challenges.

Tunisia is a good example to see how natural resources shape and affect the socio-economic and ecological systems. The country was the scene of a popular revolution in 2010-2011 against dictatorship and social inequality. The wind of revolution started in the mining region and neighboring cities in the first attempt in 2008. The movement was oppressed by police violence against the local population. In late 2010, a more fruitful and powerful popular movement started again in the same region and spread nationwide to led to the revolution of January 2011. The population aspired for rights like employment and access to health care, and social equity between regions. Since the Tunisian revolution in 2011, the extractive sector's governance has featured prominently in the discussions about reforms in the country. This chapter is a descriptive introduction of the studied area and the phosphate mining activity in the region.

5.2 General description of the mining region of Gafsa

The mining area is located in the governorate of Gafsa (see Figure 5.1). Gafsa has 13 administrative municipalities or cities [147]. The mining area is composed mainly of four cities: Metlaoui, Moulares, Redeyef, and M'dhila.

5.2.1 Historical background of Phosphate mining in the Gafsa region

April 1885, Philippe Thomas veterinarian of the French army and geologist, discovered powerful layers of calcium phosphate on the northern of Seldja´s Mountains (Part of Metlaoui currently) the south of Tunisia. Other geological surveys and large-scale explorations followed this discovery, which revealed important phosphate rock deposits in southern Tunisia [7] [15].

Figure 5.1: Location map of CPG's mining sites and its beneficiation plants in Gafsa region Tunisia [35]

The industrial exploitation of phosphate began in 1897 with a concession granted to the Phosphate and Railway Company of Gafsa (Compagnie des phosphates et Chemin de Fer de Gafsa) [15]. A 150 km railway line between Metlaoui and the harbor city Sfax allows the minerals to be transported to the port and then exported. This railway network has also linked all the surrounding mining villages [18].

The traditional exploitation of underground mines was extracting the minerals by digging tunnels or galleries up to 300 meters deep. The low-powered phosphate rock layers were blasted, and the product was picked up by shovels in the small sedans. The traction of the wagons was ensured by animal power.

Metlaoui, the first mining city, was a village of indigenous people (tribes). Under the French protectorate, it experienced substantial economic and social prosperity thanks to the development of the mining activity. The genesis of the different mining cities is closely linked to the development of the phosphate rock mines by attracting the flow of immigration [18] [173]. The European workers (French and Italian) were occupying the administration and leading position, and the Algerian, Libyan, Moroccan, and Tunisian workers were the miners [173].

This situation changed gradually after the independence of Tunisia in 1956. The region experienced a metamorphosis of its urban morphology. Between 1940 and 1970, anarchic urban clusters were built and extended during the 80s around the industrial area [173]. The continuous extension of both industrial and urban areas in the same space led to a chaotic spatial structure of mining cities (in some cases, the conveyor belt crossed the city-center where some houses and shops are positioned directly under it).

Moreover, the change in the way phosphate rock was extracted from "underground" to "open-pit" and the mechanization of phosphate extraction during the period 1980-1987 led to early retirement or arbitrary expulsion of about 6,000 workers [173]. This event negatively affected the operational capacity of the region, resulting in animosity and aversion towards the Company of Phosphate of Gafsa (CPG) [143]. Since then, the recruitment of miners has become the primary source of conflict, thus, affecting the relationship between local communities and the government of the region.

5.2.2 Geological and climate description

The Gafsa phosphate mining region, known as the Gafsa mining basin, is located in the south-west Tunisian Saharan Atlas [94], between the High Steppes and the Sahara [173]. This basin presents a specific climate pattern and currently covers an area of 3,000 km^2

[156]. It is home to 107,928 inhabitants, which represents 32 % of the population of the governorate of Gafsa (337,331 inhabitants) in 2014 [147]. Its geological formation corresponds to a synclinal structure limited in the north and the east by the mountainside, in the south by the northern Chott range, and in the west by the Algerian territory. The formation consists of fold belts caused by the collision of the Eurasian and African plate [94].

- **The genesis of phosphate formation**

The majority of the mined deposits date to the Paleocene (66-56 million years ago)-Eocene age (56 -33.9 million years ago) [121]. Before being the Sahara, the region was covered by the sea. During this period, the sea experienced a proliferation of biomass (phytoplankton and zooplankton, fish, shellfish, and marine plants) favored by a significant frequency of "upwelling" marine currents, which are nourishing currents, bringing dissolved phosphorus and other nutrients, to where the marine fauna and flora is concentrated [163] [121]. During the late Paleocene (Lower Ypresian), sudden variations in water temperature, pH, and salinity occurred and caused the mass death of marine life [163]. After the sea's withdrawal from the region, the degraded organic substances generated the formation of ammonium carbonate and calcium phosphate.

- **Climate description**

The mining area of Gafsa has a Mediterranean climate with arid and semi-arid variants. It is typically characterized by a significant annual variation of temperature. The cold season lasts for 3.6 months, from 20th November to 7th Mars, with an average temperature below 17°C. Statistically, the coldest day is the 11th of January, with an average minimum temperature of 4°C in Redeyef and 6°C in Metlaoui [221]. The summer lasts for around three months, from the 7th of June to the 11th of September. The temperature varies between an average minimum of 25°C and an average maximum of 38°C. Statistically, the hottest day is the 31st of July, with an average maximum temperature of 38 °C [221]. The temperature may exceed 40 °C during 19 to 30 days per year [147].

The studied area is subject to the north-northeast wind regime throughout the year. The south wind becomes frequent only in spring and summer and shows a maximum wind speed of 24 $m.s^{-1}$ [147]. Hot winds blowing from the southeast could take place with speed higher than 6 m/s [147]. It often brings excessively high temperatures and unpleasant dust that lingers in the air and covers everything. It is publicly known as "Sirocco" (or in the local population as Chehili).

For the precipitation, the average annual rainfall is estimated to 203.7 mm per year [147]. The wet season lasts six months from the 1st of November to 1st of Mai, with an average monthly precipitation of 17 mm accumulated in January [221]. The dry season lasts six months from the 1st of Mai to the 1st of November, with an average monthly precipitation of 1 mm in July [221].

- **Water resources**

According to Salhi (2017), there are four surface water tables in the mining area. The surface water resources are evaluated to 8.47 Mm^3/year with an exploitation level of more than 9 Mm^3/year. The exploitation rate is about 110% between 2012 and 2016 [173]. The groundwater resources are estimated at 17.8 Mm^3/year, exploited by 25 drilling-wells [173]. According to the Regional Office of the Development of the South (ODS) (2016), the surface water resources are 33.3 Mm^3/year in the total governorate of Gafsa (the totality of 13 cities), and the exploitation is 51.73 Mm^3/year. The regional office estimates the total groundwater resources in the Gafsa governorate to 95.10 Mm^3/year, and the exploitation is estimated to 97.56 Mm^3/year [147].

The salinity of the groundwater is highly variable, ranging from 1.5 g/l to 11.5 g/l [173]. The water boreholes, which are located in areas of a high salinity of 11.58 g/l, have been allocated to meet the industrial water needs of the phosphate industry in Moulares and Redeyef (because of their low chemical quality) [173].

Both surface and groundwater are overexploited. The natural recharge of the water table is not only through direct infiltration of rainwater but also from the floods of the wadis [173]. Due to low rainfall and high water exploitation by agriculture and the phosphate industry, the water resources in the region are reported to be depleting [18, 173].

- **Wild fauna and flora**

The wildlife in the desert is pretty rare. However, a small mammal such as Fennec fox and Jerboa are adapted to the desert conditions [18]. The vegetation cover is composed mainly of Tamarix aphylla, Salicornia arabica, Atriplex L., and Polypogon. These plants play a role as pasture resources and to fix soil to reduce desertification. Salhi (2017) has compared the vegetation cover in 1987 and 2015 and has reported regression of the vegetation in the zones of open-pit mining extensions, especially in the area where the rejection of the beneficiation plants are transferred to the ponds [173]. The vegetation coverage index decreases remarkably around the mining cities. It is mostly the areas with pastures that are affected [173].

5.2.3 Socio-economic characterization

This section discusses the demographic and educational characteristics of the mining region's population (Moulares, M'dhila, Redeyef, Metlaoui) as well as the economic characteristics such as the labor market and the unemployment rate.

- **Population structure**

In 2014, based on information provided by the National Institute of Statistics of Tunisia (INS), over 107,000 people were living in the Gafsa mining basin, with an almost equal balance between both genders, males (53,713) and females (54,215) [147] (see Table 5.1). Around 90% of the local population lives in an urban area. The average population density in the Gafsa governorate is 43.2 habitants/km^2 [147].

Table 5.1: Local population demographic characterization [147]

	Population thousand	**Female** thousand	**Male** thousand	**Family size**	**Urbanization** %
Metlaoui	38.634	19.255	19.379	4.3	99%
M'dhila	15.306	7.524	7.782	4.7	84%
Moulares	27.012	13.684	13.328	4.4	79%
Redeyef	26.976	13.752	13.224	4.5	93%
Total/ Average	**107.928**	**54.215**	**53.713**	**4.3**	**89%**

- **Education**

The educational system in Tunisia is structured into basic (mandatory) and secondary education. The basic education starts with primary school from grades 1-6, then three years of basic education, grades 7-9, with students prepared to pursue high school or to go for professional training. The secondary education is four years of high school (grades 1-4), which cover the pre-university public schooling and end with a national exam allowing students to access university. Table 5.2 shows that most of the population drop out of secondary school. An average of 10% has a university degree in the mining region.

The average illiteracy rate in the mining regions in 2014 was 20.8 %, of which women take a significant share (see Table 5.2). The average primary school enrollment of children aged between 6-14 years old is 97%. The average enrollment at the university among the young population is 41.4%, where women are more likely to enroll at the university after high school. Furthermore, the slight difference between the ratio of female and male enrollment shows that men in the mining area tend to follow the path of professional training.

Table 5.2: Educational characteristics of the local population of the mining area in % [147]

	Primary school		High school		University		Illiteracy rate		Enrollment in school 6-14 years		Enrollment at the university 19-24 years	
	Male	Female	Male	Female	Male	Female	Male	Female	Male	Female	Male	Female
Metlaoui	32.8	32.5	48.3	35.8	10.3	11.3	8.6	18.6	97.7	97.0	43.9	52.8
M'dhila	32.0	27.5	40.8	31.6	9.1	9.4	17.9	29.5	97.0	96.7	36.0	47.2
Moulares	28.7	25.9	41.7	30.2	10.7	10.4	18.9	32.3	97.7	96.3	33.3	42.0
Redeyef	31.2	29.7	44.0	34.1	9.9	12.9	14.9	25.7	97.5	97.0	33.4	41.6
Average	**30.025**		**38.25**		**10.77**		**20.82**		**97.1**		**41.47**	

- **Unemployment**

According to the OECD, "the unemployment rate" is the number of unemployed people as a percentage of the labor force, where the latter consists of the unemployed plus those in paid or self-employment" [153]. Unemployed people are those who are able to work but reported without work, and they have been looking actively for work in the last four weeks [153].

The total unemployment rate in the mining region is relatively high. It is double the national unemployment rate (see Figure 5.2-a). The most impacted category is the graduated young people (41.5% unemployment rate). Besides, unemployment among women is 2.3 times the unemployment rate among men (43.7% vs. 19.2%) (see Table 5.3). This vast gap reflects gender iniquities and the women's vulnerability in the labor market in the mining region.

Table 5.3: Unemployment rate by gender and university graduates in the mining region in % [147]

	Women	Men	Total
Unemployment rate	43.7	19.2	26.3
Unemployment rate among graduates	54.6	27.8	41.5

According to Salhi (2017), the Company of Phosphate of Gafsa (CPG) was the major employer in the region between 1950 and 1960, employing 80% of the active population and providing affordable electricity, drinking water, and rail infrastructure [173].

This situation has changed between 1970 and 1980. The mechanization of phosphate extraction and the company restructuring led to a wave of early retirement at the Company of Phosphate of Gafsa (CPG). The number of employees, specifically the miner workers, dropped dramatically from 14,000 in the 80s to 5,300 in 2007 [85, 173]. The unemployment rate increased dramatically in the region. Despite the adoption of mechanization and the downstream production strategy to increase the economic value of exportation, the regional policy in the 80s did neglect the employment opportunities in the mining region. The administrative sector and agriculture sector could not meet the need for employment in the region.

In 2004, the phosphate industry assured around 20% of the employment of the active local population in comparison to more than 70% in 1975 (see Figure 5.2-b). The local unemployment rate increased to 25% in 2008 and reached 28% in 2010, which is double the national rate (see Figure 5.2-a).

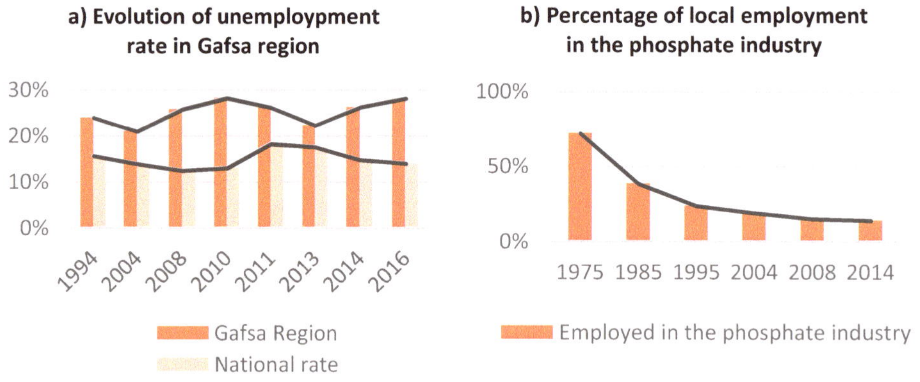

Figure 5.2: Comparison of the evolution of local unemployment with the active population in the phosphate industry in Gafsa, based on data from INS and ODS [52, 102–104, 107, 147, 149]

- **Labor market**

The employed population is mainly active in the service sector and the mining sector, with a rate of 60% and 29% respectively in 2014 [147]. It reflects the predominance of the public sector in the local economy and the role of the phosphate industry as a significant employer in the region. According to Table 5.4, employment in different economic activities is not equally distributed among the sectors. According to OECD (2015), economic activity is highly concentrated in the interior region, with very limited private companies investing in these regions [151]. In contrast, the northern and eastern regions, along with the coast, host 74% of the private enterprises accounted in the National Registry [151]. The problem of economic activity concentration is not only related to the regional disparity of the presence of private enterprises but also because the few enterprises existing in the region tend to operate in the same sectors, increasing output volatility and making these regions more vulnerable to shocks [151]. The market concentration index expressed by the Herfindahl-Hirschman Index (HHI)[1] in the Gafsa governorate is 0.3, which is very high (>0.18) (see chapter 2 section 2.3.1) [151].

Table 5.4: Employed population by sectors expressed in percentage of the total employed population [147]

Sector	Agriculture	Manufacturing	Mining	Construction and public civil work	Service	Total employed population
Metlaoui	1%	4%	35%	7%	53%	8,258
M'dhila	1%	8%	32%	4%	55%	3,807
Moulares	2%	2%	24%	5%	67%	5,166
Redeyef	1%	2%	22%	3%	70%	5,028
Total	**1%**	**4%**	**29%**	**5%**	**60%**	**22,259**

5.2.4 Socio-political situation

The discovery of the phosphate deposits has brought profound socio-economic changes. Worth repeating that the development of the mining industry and economic activities has

[1] For details about the HHI see chapter 2 section 2.3.1

been followed by considerable development of infrastructures. The mining community has expanded, bringing together a large society with different traditions and cultures [18]. However, this positive impact did not last too long with the appearance of environmental and socio-economic disparities between regions, mainly between the regions along the coast and the interior regions. The phosphate industry has been considered as 'an initiator of life and initiator of decay' [125]. Hence, the importance to report the socio-political history of the mining region.

The mining region has been facing injustice and oppression for a long time. Starting in the French colonial era and followed by successive regimes, this made the region a hotspot for revolutionary attitude [143]. The story began over a half-century ago when an uprising against the president Habib Bourguiba[2] orchestrated by Gafsa citizens over the president's oppressive measures [172]. On 26th January 1980, a group of militants seized police and military installations in the southern phosphate mining center of Gafsa to spark a general uprising against Bourguiba's oppressive neoliberal policies, which ended by the arrest of hundreds of unionists.

In early 2008, a social movement started in the mining region of Gafsa, labeled as "The Gafsa Mining Basin Revolution" [85] in the city of Redeyef and was ended by violent police oppression of the local population. According to Gobe (2011), it was the most exhaustive mass actions since the Bread riots[3] in 1984. The flashpoint was when the Company of Phosphate of Gafsa (CPG) announced a hiring contest for workers and offered job opportunities to candidates who were not from the local population [172]. The local population believed that the hiring process had been manipulated.

On 6th January 2009, a small group of trade union activists demanded from the state to set up a negotiation committee to discuss regional issues. By the end of the same year, the trade union leaders have been sentenced to 10 years in prison [172]. The movement was suppressed, but Gafsa became a central player in the revolution against Ben Ali[4]'s authoritarian regime [172].

---

[2] Bourguiba was Tunisias first president after gain of independence in 1956-1987. He abolished the royal regime and proclaimed Tunisia as republic in 1957.

[3] The bread riots started in southern Tunisia and spread to a violent demonstrations to the country's two largest cities, Tunis and Sfax over food- price increases [189].

[4] The second Tunisian president after the independence of Tunisia. He rose to the presidency of the country on 7th November 1987 in a bloodless coup d'état that ousted President Habib Bourguiba by

In 2010, a second popular movement started again in the mining region of Gafsa and led to a Tunisian protest across the country, targeting the system's corruption and nepotism. The situation evolved quickly from local problems to a national revolution and economic crisis [8], as indicated in Figure 5.3.

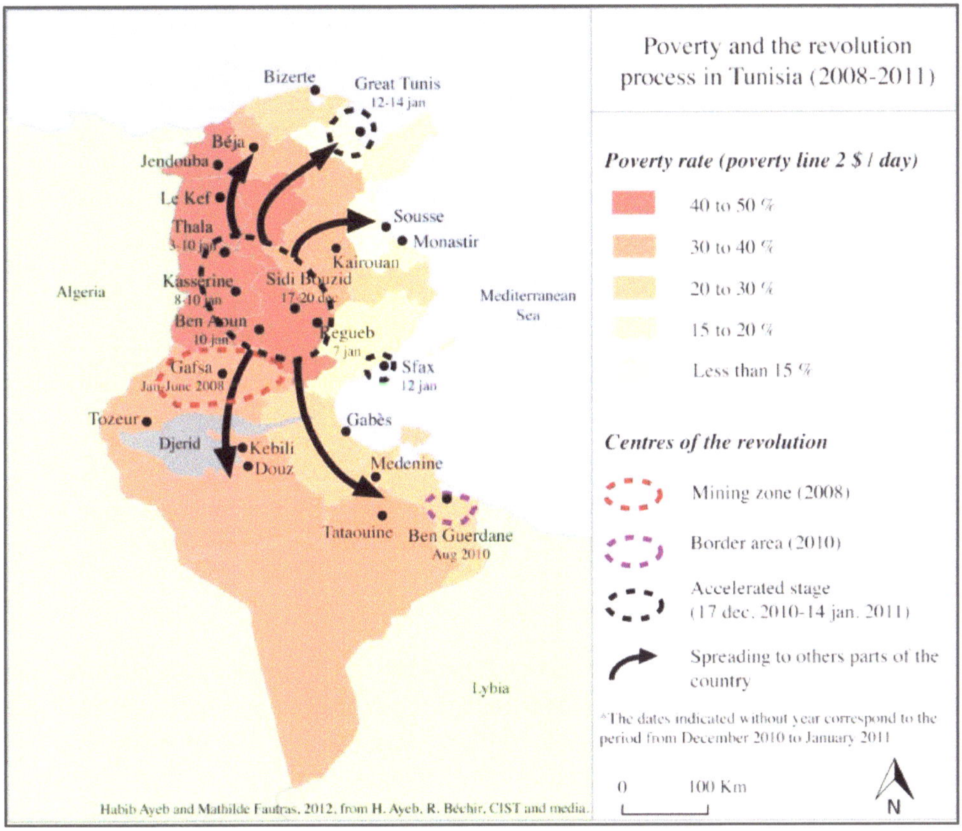

Figure 5.3: Evolution of Protests in Tunisia from 2008 to the 2011 Revolution [8]

Protestors have undercut the company's performance. Camps were set up in front of the beneficiation plants and blocked the railway to stop the trains carrying beneficiated phosphate rock. At the same time, young unemployed men set out to tear apart railway tracks

declaring him incompetent [190]. He ruled the country until his fall in 2011. His regime was known as a police state. He was forced to leave the country on the 14th of January 2011 threatened by the population revolution.

linking M'dhila to Moulares. Since then, natural resources governance has featured prominently in discussions about reforms in the country [98, 180].

The social and economic factors behind the unstable situation in the mining region can be summarized as follow; first, the high unemployment rate, second the feeling of being neglected by the state and excluded from the development strategy, and third the high disparity in local investment between the coastline region and the interior region. Besides, the local population considers themselves suffering from the negative environmental effects of the mining industry. They do not benefit from the economic spin-offs of the industry, and they do not have a community right to compensation for environmental and health damages.

5.3 The Company of Phosphate of Gafsa (CPG)

5.3.1 Presentation

The Company of Phosphate of Gafsa (CPG) is a state-owned enterprise. It has been active in phosphate mining for more than a century since its creation in 1897. In 2016, it was counting 6,619 employees.

It ranks 5th among ten largest producers of beneficiated phosphate rock worldwide in 2010 with an annual production of 8 MMT before the drop of the production in 2011 (2.2 MMT). In 2017, the company produced 3.9 MMT of beneficiated phosphate rock.

After having been exporting all its phosphate rock production during fifty years of its activity, Tunisia developed the phosphoric acid and mineral fertilizer production industry (downstream production strategy). Currently, more than 80% of the beneficiated phosphate rock production is being processed into phosphoric acid and fertilizers [21]. The phosphate chemical industry mainly consists of two companies; the state-owned company "Groupe Chimique Tunisien" (GCT), which delivers phosphoric acid and phosphoric fertilizers, and the "Tunisian Indian Fertilizers company" (TIFERT), which specialized in phosphoric acid production for the Indian market. The mining company CPG and GCT merged 1996 to form the CPG-GCT group.

CPG key dates since the discovery of phosphate in Tunisia [37]:

1885: Discovery of phosphate deposits by Philippe Thomas on the northern slope of Jbel Thelja near Metlaoui.

1897: Foundation of the *"Company of Phosphates and rails of Gafsa"* (CPCFG) and the beginning of the first extraction site in Metlaoui. At the same time, the start of the construction of the railway line linking Metlaoui to the port of Sfax.

1899: Opening of the first underground mine (at Metlaoui)

1905: Foundation of the *"Phosphate Mining Company"* (STEPHOS)

1920: Foundation of the *"Tunisian Company of Phosphates de Jbel M'dhilla"* (STPJM)

1969: Merger of the STPJM and CPCFG

1976: Merger of STEPHOS and the CPCFG; starting from that date, all the companies come together to form the *"Phosphate company of Gafsa"* (Compagnie des Phosphates de Gafsa – CPG)

1978: Operation of the first surface mine (Kef Schfaier) ad creation of the Research Center in Metlaoui

1996: Merger of the two commercial structures of CPG and GCT to form the CPG-GCT group

2006: Closure of the last underground mine (Redeyef)

5.3.2 Spatial distribution of Phosphate activities in the Gafsa region

The company operates five production sectors with nine mining sites for phosphate extraction and ten washing plants for the production of the marketable beneficiated phosphate rock (see Table 5.5). The sectors extend on a distance of 80 kilometers from M'dhilla (east side) to Redeyef (west side), going through Metlaoui and Moulares [35]. According to the Research Center report, the phosphate deposits in Gafsa are estimated to cover a total surface of 5,000 to 6,000 km [163]. The current mining area covers a surface of approximately 200 km^2, according to the Google Map Area Calculation Tool [5] (see Figure 5.4).

Table 5.5: Mining site organized by sector according to the administrative division of the company

Sector	Metlaoui-kef Schfaier	Metlaoui-kef Eddour	Redeyef	Moulares	M'dhila
Site	Tables de Metlaoui	Kef Eddour centre	Redeyef	Moulares	Jellabia
	Kef Schfaier	Kef Eddour ouest	-	Om Lakchab	Mzinda
Beneficiation plants	4 plants	1 plant	1 plant	1 plant	3 plants

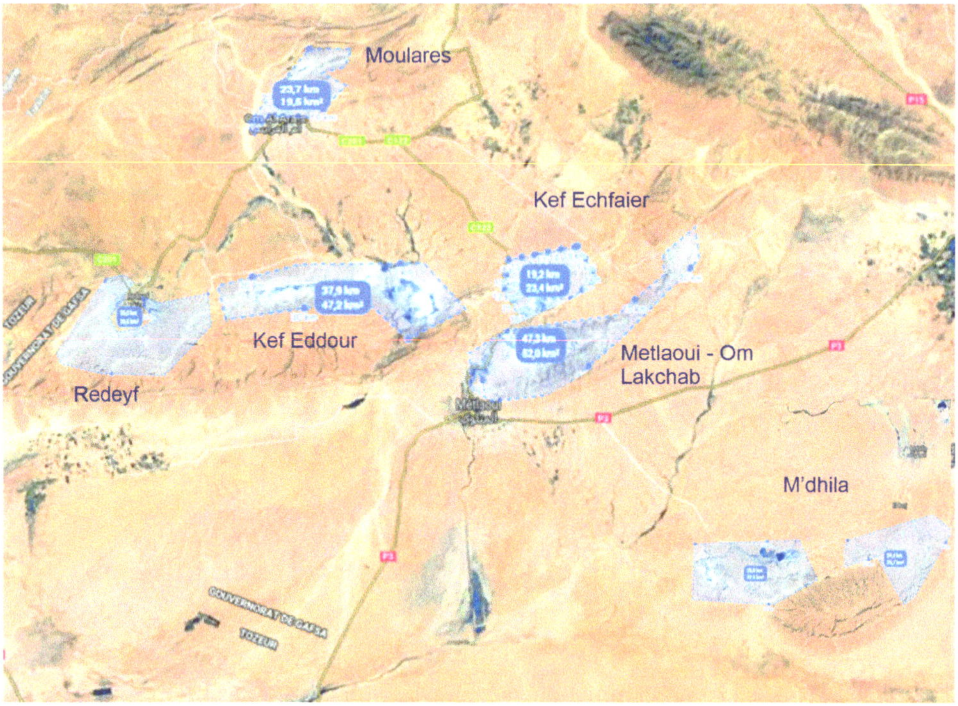

Figure 5.4: Estimation of the current surface of mining sectors in the mining basin of Gafsa according to the Google Map Area Calculation Tool [5]

5.4 Conclusion: Challenges of the mining region

Local communities are inherently affected by a detrimental environmental, social, and economic situation. The environmental challenges are mainly water scarcity creating a deficit for agriculture and farming, and the environmental pollution reducing the region's livelihood and affecting the population's health status. The social and economic challenges are mainly the unemployment rate and the high disparity between regions towards private investments and the creation of jobs.

As the CPG is a state-owned company and its history in the region of being the company, which produced and distributed electricity and water and affordable transportation for the local population in the past – the local population considers the company management is representing the state and its regime. After the revolution, the CPG found itself in a dilemma. Everyone wants to have a job at CPG, but at the same time, they point the blame finger to the company because of the environmental and socio-economic problems in the region. On the one hand, the company is accountable to the state to deliver a positive image to the international partners regarding political stability and good governance practices and keep its international market share. On the other hand, the company is accountable to the society and local population to foster local socio-economic development and improve its environmental performance.

This imposes the challenge of how the company can continue to operate in such a turbulent social environment. The company is aware of the necessity to take active actions to regain the local population's trust in order to restore its production capacity. Therefore, an integrated environmental and social impact assessment was required by the company. Such an assessment would help to identify the socio-economic linkages to environmental issues.

6 Environmental life cycle analysis of phosphate rock mining

6.1 Introduction

In this chapter, an environmental Life Cycle Assessment will be conducted for the production system of beneficiated phosphate rock in the mining region of Gafsa. The structure of the chapter follows the four steps of the LCA Methodology. The first step sets the goal of the study, defines the investigated environmental issues, and determines the limits of the product system. Then, in the inventory analysis, the material flows through the studied product system are quantified. After defining the input-output flows, the environmental impacts are assessed according to the selected LCIA method. The study is concluded by an interpretation of the results and recommendations.

6.2 Goal study

The goal of this study is to attribute an environmental performance to the production of beneficiated phosphate rock in Tunisia, to assess the environmental impacts concerning all the emissions to air, soil, and water, and to determine the potential impacts of phosphate activities on human health.

The results of this study will be used for the discussion of the interconnection between environmental impacts and social impacts of phosphate mining and their effect as a supply risk.

6.3 Scope

6.3.1 Product system definition

The considered product system is the production of beneficiated phosphate rock in the mining basin of Gafsa in Tunisia from cradle-to-gate (see Figure 6.1). The scope of the analysis is referring to an average production capacity of 5 MMT of beneficiated phosphate rock per year.

6.3.2 Functional Unit

Phosphate rock is the primary source of phosphorus, which is used in many industrial applications. The main application of phosphate rock is the production of phosphorus fertilizers. Phosphorus pentoxide (P_2O_5) is used as a reference to express the phosphorus content in the commercial fertilizers and phosphoric acid (H_3PO_4). The commercial beneficiated phosphate rock has to contain at least 28% of P_2O_5 to produce a fertilizer [159]. Besides, the geological grade of phosphate rock refers to the P_2O_5 content of the crude ore. For this study, the functional unit is 1 kg of Phosphorus Pentoxide (P_2O_5).

The mineralogical composition of the Gafsa basin phosphate is heterogeneous, with quantitative variations of the P_2O_5 content between the deposits and also between different layers in the same deposit. Besides the P_2O_5 content, other chemical elements can vary. For this inventory, the average content of the crude phosphate rock in Tunisia is 24% P_2O_5, and the average content of the beneficiated phosphate rock is 29% P_2O_5. Table 6.1 shows the average chemical properties of Tunisian beneficiated phosphate rock.

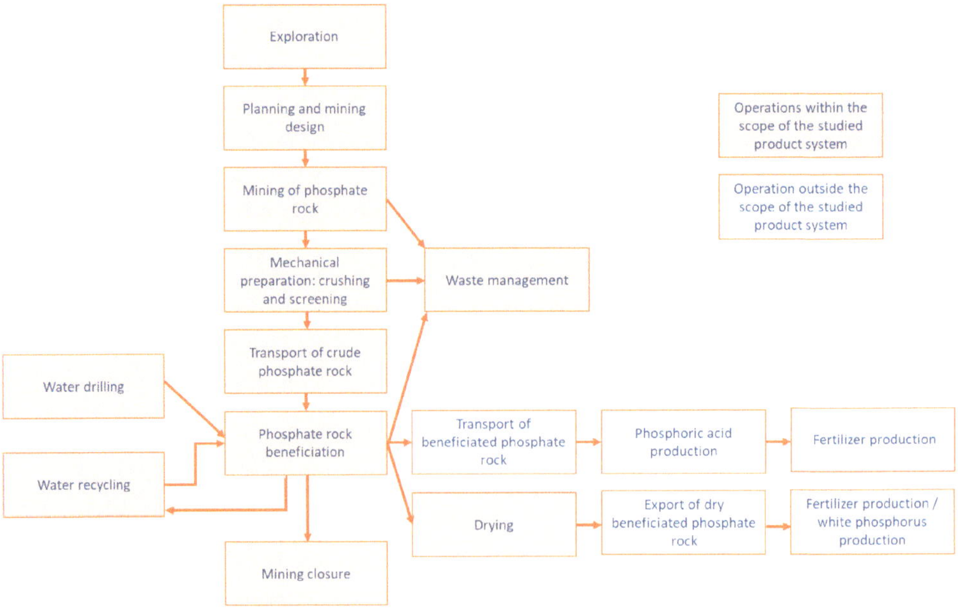

Figure 6.1: Product system and the scope of the environmental Life Cycle Assessment

Table 6.1: Chemical properties of beneficiated phosphate rock in Tunisia according to the chemical analysis at the CPG [164]

Element	Unit	Average Value [a]	CV[b]
Phosphorus as P_2O_5	%	29	2%
Calcium Oxide CaO	%	48.33	1%
Magnesium Oxide MgO	%	0.52	14%
Carbon organic	%	0.75	24%
Cadmium Cd	ppm	36.7	36%
Uranium[c] U	ppm	46.5	-
Thorium[c] Th	ppm	10.20	-

a: The average of 9 mining sites (the value can change for one mining cite as the sample composition from different layers can change)

b: The Coefficient of variation

c: The value of U and Th are taken from the doctoral work on the radioactive element in Gafsa phosphate at the CPG [78]

6.3.3 System boundary

The system boundary for the production of 1 kg of P_2O_5 defines the space where inputs are processed and transformed into the final product and thereby generate outputs and wastes. The final product of the evaluated product system is the beneficiated phosphate rock, concentrated at 29% P_2O_5, the marketable product. The functional unit, which is 1 kg of P_2O_5, is equal to 3.45 kg of wet beneficiated phosphate rock.

The wet beneficiated phosphate rock is transported to the phosphoric acid plants in different locations in the country. Drying concerns only the exported product. The average production of dry beneficiated phosphate rock is 2% of the total produced beneficiated phosphate rock per year. A simplified process representation of the beneficiated phosphate rock production in Tunisia is shown in Figure 6.1. A detailed description of each process unit is carried out in the LCI section.

6.3.4 Preparing the basis for the impact assessment

The selected impact categories from the applied ReCiPe Midpoint method (see Annex D) are related to three areas of damage indicated in the goal of the study (the full list of impact categories and characterization factors in ReCiPe can be found in Annex B, Table B-1). The evaluated damaged areas caused by the industry are damage to water, damage to human health, and damage to arable land. The environmental impact categories are presented in Figure 6.2.

Normalization and weighting were excluded from the study. It is considered to be an optional step for the LCIA [74]. Normalization aims to improve the interpretation of the impact by expressing midpoint level impact categories relative to a common reference value. The reference value is usually the impact of the total annual emission or resource use per capita in a country, region, or globally. However, the ReCipe method has two sets of normalization either for the year 2000 or per person in Europe and the World. It has been assumed that the normalization set in ReCiPe is outdated (data for normalization inventories are twenty years old). Moreover, the goal of the study does not stipulate a comparison to the global (or European) average impact per year per person. The interpretation of the LCIA must quantify the environmental impacts related to the studied region conditions and context.

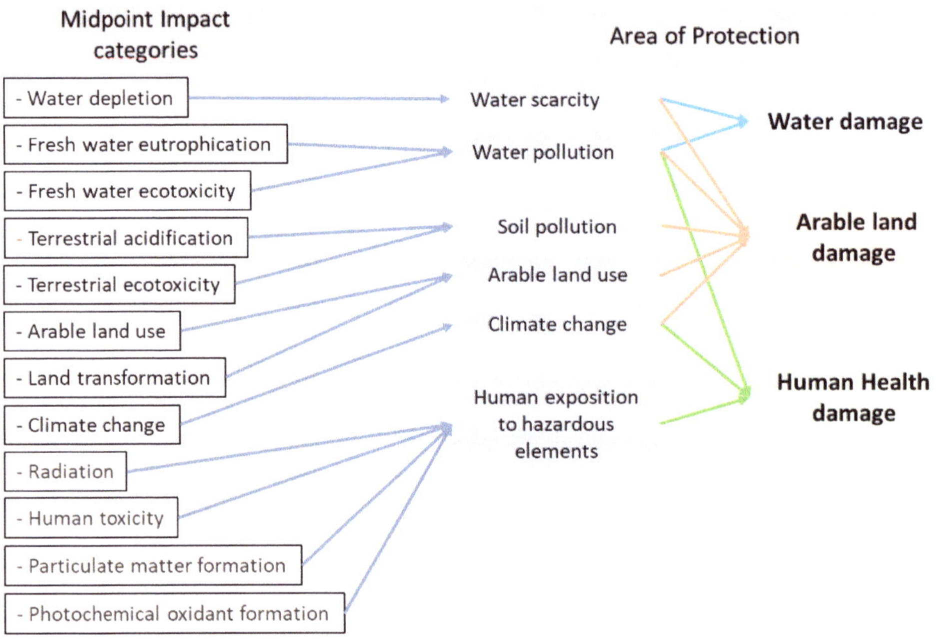

Figure 6.2: Selected environmental midpoint impact categories for the studied area of damage

6.3.5 Type and source of data

The data selected for a site-specific LCA (type "cradle to gate") were collected from the nine mining sites for the mining operations, from the ten washing plants for the wet beneficiation process, and from three drying units for the drying process. The water recycling process inventory is included in the wet beneficiation process inventory, as each washing plant has a water recycling station. The data were made available by the company CPG.

The collected data are mainly inputs (chemicals, energy, water, fuel, phosphate rock, and machinery). The emissions were either calculated or collected from previous research projects at the research center of the CPG. The collected data are for the years 2016 and 2017. The sources of data are the official reports of the company and the material balance reports of washing plants and mining sites. For the background processes[1], generic data were used from the "Ecoinvent" database Version 3.4.

6.4 Life cycle inventory

The evaluated system covers three unit processes:

- The open-pit mining, expressed as "Mining operations phosphate rock, as P_2O_5-TN"
- The wet beneficiation of crude phosphate rock, expressed as the "Wet beneficiation phosphate rock, as P_2O_5-TN"
- The drying of the beneficiated phosphate rock, expressed as the "Drying phosphate rock, as P_2O_5-TN"

A general overview of the total inputs used in different processes per year is presented in Table 6.2. The detailed input/outputs data per functional unit (1 kg P_2O_5) are in Annex D, Table D-7.

[1] The background system consists of processes on which no or, at best, indirect influence may be exercised by the decision-maker for which an LCA is carried out. Such processes are called "background processes." [81].

Table 6.2: The average annual production of beneficiated phosphate rock and the corresponding annual inputs per process stages: the average of 2016-2017

Resource used	Unit	Total CPG	Mining operations	Wet beneficiation	Drying
Total Production PR[a]	kg	3.65E+09	4.89E+09	3.46E+09	1.90E+08
Land use per year[b]	m^2	1.20E+06	1.20E+06	5.31E+02	n.a
Moved soil	kg	7.88E+10	7.88E+10	n.a	n.a
Diesel	MJ	8.63E+08	8.50E+08	1.26E+07	n.a
Heavy fuel n°2	MJ	1.32E+08	n.a	n.a	1.32E+08
Electricity	kWh	4.02E+07	8.08E+05	3.94E+07	n.a
Total Energy	**MJ**	**1.14E+09**	**8.53E+08**	**1.55E+08**	**1.32E+08**
Transport payload distance	kg-km	1.99E+11	1.81E+11	1.78E+10	n.a
Explosive	kg	1.08E+07	n.a	n.a	n.a
Drilled water	m^3	1.18E+07	n.a	1.18E+07	n.a
Flocculants	kg	2.53E+05	n.a	2.53E+05	n.a

a: the average production of beneficiated phosphate rock wet+dry (2016-2017) the crude phosphate rock produced during the mining operations is an input for the wet beneficiation process

b: Land use per year is calculated based on the occupied surface divided by 20 years of operation

The functional unit 1 kg P_2O_5 is equivalent to 3.45 kg of beneficiated phosphate rock (29% P_2O_5) and 4.17 kg of crude phosphate rock (24% P_2O_5). However, it needs 5.32 kg of crude phosphate rock of 40 mm size (24% P_2O_5) to produce 1 kg P_2O_5, with an average loss of 0.3 kg P_2O_5 per 1 kg P_2O_5 during mining operations and the beneficiation according to the actual beneficiation rate (65%) (see Figure 6.3). The losses during mining operations are not considered in the output as they are regarded as future reserves.

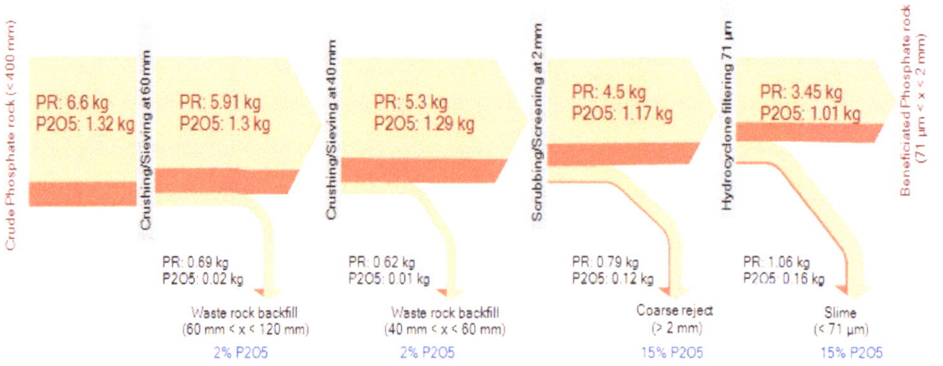

Figure 6.3: Sankey flow diagram of phosphate rock (PR) and P_2O_5 during mining operations and wet beneficiation per functional unit of 1 kg P_2O_5

6.4.1 Mining operations phosphate rock, as P_2O_5-TN

6.4.1.1 Process description

The phosphate rock is mined in an open-pit site. In the Gafsa phosphate rock deposit, there are eight layers of phosphate rock. The first and second layers represent around 50% of the phosphate deposit in a non-exploited deposit [46].

A simplified process flow of phosphate rock mining operations in Tunisia is shown in Figure 6.4. The strip ratio – the total kg moved rock per kg crude phosphate rock (40 mm size) – is very high. The strip ratio was 13:1 in 2016 [53, 60], which means for every 1 kg crude phosphate rock (40 mm size), 14 kg of soil have to be moved, and 13 kg of overburden and intermediate layers were backfilled.

The strip ratio varies from 8 to 15 in the Gafsa mining basin [60]. The overburden limestone layer is very thick, and its average height is 45 m (it can vary between 20 m-60 m). The intermediate layers vary between 50 cm and 4.5 m [46]. Consequently, the overburden layer and the intermediate layers have to be removed before mining the phosphate rock layers.

Two options are available to remove the overburden and the intermediate layers, depending on the geophysical properties of the layer. The first option is blast hole drilling using a rotary driller [60]. Then the overburden is stripped by a 15 m high bench[2] [46, 163].

The second option is blasting the overburden using the ANFO[3] explosives [60]. Blasting the overburden is only economically feasible when the overburden is very thick or the thickness of the intermediate layers is above 2 m, which is difficult for the bulldozers to take it down. After the explosion, bulldozers and small shovels extract the phosphates layers [46].

[2] A bench may be defined as a ledge that forms a single level of operation above which mineral or waste materials are mined back to a bench face. The mineral or waste is removed in successive layers, each of which is a bench [135]. http://www.mine-engineer.com/mining/open_pit.htm

[3] ANFO: **A**mmonium **N**itrate **F**uel **O**il. It is composed of 94% ammonium nitrate (NH_4NO_3) and 6% fuel oil. These explosives belong to the family of nitrated explosives

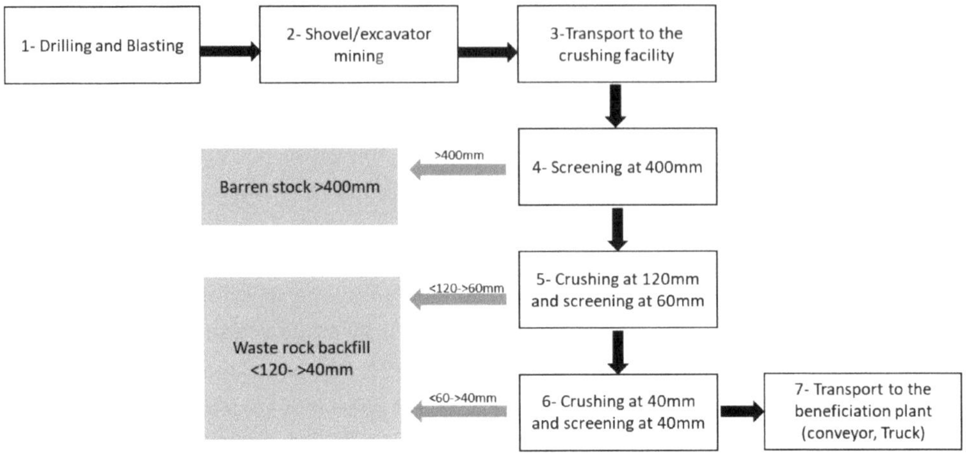

Figure 6.4: Simplified process flow of the open-pit mining based on data from the direction of mining pacification at CPG [58]

The extracted phosphate rock[4] is transported by truck (85 t capacity) and discharged in the crushing facility. At first, the rocks are sorted atop a static sieve with a mesh size of 400 mm opening in order to remove the oversized rocks (> 400 mm) [58]. The company names the oversized rock stock as "Barren rock", which refers to the negligible content or absence of phosphate rock in this fraction. The barren rocks are loaded in the dumper trucks and backfilled to the side of the mining area. Then, the rocks smaller than 400 mm are crushed in a "Jaw Crusher unit" to reach the size of 0-120 mm and sieved through a vibrating sieve of 60 mm opening mesh. During this operation, 10% is backfilled. The 0-60 mm size rocks undergo a second crushing, and then they are sieved through a vibrating sieve of 40 mm opening mesh to reach the size of 0-40 mm. During this second crushing and sieving, an additional 10% is backfilled. The sorted rocks of 0-40 mm are transported by a conveyor belt or truck to be stocked in an area next to the washing plant [58]. The backfilled rocks with a size between <120->40 mm are stocked aside [58]. This stock contains an average of 6% crude phosphate rock due to the imperfection of the process, and it can be used in the future [46].

[4] It refers to the layers of phosphate rock before to undergoes any mechanical transformation to remove the non-phosphate rock

The Tunisian phosphate is known to be friable. Around 10% of crude phosphate rock[5] is lost during the first screening operation (at 60 mm), and an additional 10% of crude phosphate rock is lost during the second screening (at 40 mm) [46] (see Figure 6.3).

6.4.1.2 Inputs of the process mining operations phosphate rock, as P_2O_5-TN

The primary operative resources are the natural resources from the ground, the machinery to perform the extraction and the transport, and the energy to run the machines. Table 6.3 displays the inputs and their flow names per 1 kg crude phosphate rock (<40 mm) and per functional unit of 1 kg P_2O_5 for the inventory of the process "Mining operations phosphate rock, as P_2O_5-TN".

- **Land use and occupation**

For the process flow "Occupation mineral site", the land occupation was calculated based on the planning map of the mining site "Metlaoui" (see Annex D, Figure D-2) for an average occupation period of 20 years. The average total occupied area per year is 1.2 km^2 to extract 4,082 kg of crude phosphate rock per m^2 [62]. The overburdens are stockpiled in an area that covers 50-60% of the total occupied area. The large surface for the backfilling of the overburden is explained by the fact that the ratio of overburdens and intermediate layers to the phosphate rock layers is very high and varies between 8:1 to 15:1 [60][60][37].

For the process flow "land transformation", the transformation of the land from desert to mineral extraction area is selected. After the reclamation of the mining area, the overburden and settling ponds (tailing of the washing plants) are considered as a future reserve of phosphate as they contain a fraction of phosphate rock, which is currently not economically feasible to recover. Therefore, the area is transformed back to the mineral extraction area.

The land transformation area is around 200 km^2, estimated according to the satellite pictures, and approved by the CPG (see Figure 5.4) [5]. For the process flow "Occupation, construction site", the occupation time was estimated to be five years. It is the time needed for building and removal of the facilities after closure [159].

[5] It refers to the phosphate rock after the crushing and screening to 0-40 mm and before the wet beneficiation

- **Land use and occupation**

According to the mineralogy report of the research center at the CPG, the main constituent of the phosphate rock is apatite [163]. In the Gafsa phosphate basin, the most frequent mineral is the Carbonate Fluorapatite or Francolite; its chemical formula is Ca10 (PO4)6 F2 [163]. The major elements in the phosphate rock are Phosphorus Pentoxide (P_2O_5), and Carbonate (CaO), which are related to the presence of the Francolite and the carbonates.

Silicon Dioxide (SiO2) is related to the presence of Quartz, Zeolites, and clays, and Magnesium Oxide (MgO) is related to the presence of dolomite and clays. The Iron (Fe) and Aluminum (Al) are bound to clays, Fluorine (F) is strictly related to the presence of Francolite. Besides, the phosphate rock contains trace elements such Sr, Ba, V, Ni, Cr, Zn, Cu, Mo, Se, Cd, U, Th, and rare-earth elements [163]. The concentration of trace elements is highly variable from one deposit to another and from one layer to another in the same deposit and is measured in ppm [163].

For this inventory, resources from the ground are Phosphorus (P) and Fluorine (F). Fluorine was considered because of the production of Fluosilicic acid during the phosphoric acid production. Other resources, such as SiO2 and Mg, were not considered in this inventory because they are not economically feasible to extract or recover. The same applies to the overburden and intermediate layers. They were not considered as input because as they are backfilled into the mine.

The Tunisian phosphate contains an average of 11% P in crude ore (18% in apatite) and 2.4% of F in crude ore (4.5% in apatite) (calculated based on an average of 24% of P_2O_5 content of the crude phosphate rock, see Annex D, Table D-1). In other words, 1 kg of crude phosphate rock contains 0.11 kg of P and 0.024 kg of F, and the functional unit 1 kg of P_2O_5 contains 0.456 kg of P and 0.10 kg of F.

- **Energy**

The mining operations, such as drilling and excavation, use diesel driven engines. Trucks of 85 metric tons are used for the transport of phosphate rock, and Dumpers used for the transportation of the overburden. The diesel is supplied by the National Company of Oil Distribution (SNDP). Electrical energy is used for the crushing and screening and the transport by conveyor [53]. The conveyor is used when the distance between the crushing unit and the washing plants is not superior to 10 km. The medium voltage electric energy is supplied by the Tunisian Company of Electricity and Gas (STEG) [53, 183].

The mining operations require an average of 0.174 MJ final energy[6] per kg of crude phosphate rock (0.725 MJ per kg P_2O_5) as diesel burned in the machinery [53, 54]. For the process flow, "Diesel burned in machinery", the machine production process was excluded as the machinery was included in the process of different heavy industrial machines (see Table 6.3). The electricity used during the mining operations is marginal (1.65E-04 kWh per kg of crude phosphate rock and 2.5E-03 per kg P_2O_5) [53, 54]. The process flow used was the "Electricity, medium voltage, TN".

- **Infrastructure and Transport**

The list and equipment used during mining operations were taken from mining sites inventories [60] (see Annex D, Table D-2). The value of their weight and characteristic were taken from the manufacturer's technical sheets. The duration of use was assumed to be 20 years for a machine as some machines are in operation since 2000 (excluding the lifetime of tires and other reparation) [60]. For the transport operation, the list of the trucks and shovels was taken from the company (STTPM[7]) [184].

The infrastructure consists of the storage building, the administration offices, the maintenance facilities, the training center, and the research center. For the process flow of "Building hall occupation", the buildings' surfaces were estimated from the satellite photos from google maps (see Annex D, Figure D-3). The lifetime was assumed to be 25 years, as suggested by the Ecoinvent Report N°.8 [159]. The average building occupation time during mining operations is 20 years, with an additional five years for the construction and the removal of the buildings. The maintenance facilities and the storage of different products (fuel, explosive) were assumed to be steel constructions (according to the visit to the maintenance facility). The process flow for the transport includes the infrastructure such as the truck itself, the maintenance, and the road.

The road transport processes were divided into transport between mining sites and transport to the mining area. The transport between mining sites is the transport of the crude phosphate rock between sites with trucks and the transport of the overburden with dumpers carried out by the STTPM [185]. The transport to the mining area for different types of equipment and products is the marine transport from the importing land and

[6] It refers to the consumed energy by the end user (the machine at the mining site) it excludes the energy used by the energy sector such as the electricity used in the oil refinery.

[7] It refers to : **S**ociété **T**unisienne de **T**ransport des **P**roduits **M**iniers. The company responsible of the transportation of the mining products. The CPG outsourced the transportation of the crude phosphate rock as well as the overburden to the STTPM.

road transport from the port to the mining area. The sea transport routes of the imported machinery were estimated using the online calculator for a logistic sea route from "www.searates.com" [178]. The road transport of the imported machinery from the port of Rades-Tunis to the assembly site at the CPG is 500 km. The transport of the explosive from the producing company SOTEMU in Gettar-Gafsa to the storage facility at the CPG is 60 km. The explosive transport is considered in the process flow "Market of Ammonium nitrate in Tunisia."

- **Explosives**

The explosive used in the blasting step is a mixture of Ammonium Nitrate fuel (ANFO) and Nitrate explosive mentioned as N30 in the reports of the mining sites [54, 60]. The ratio N30/ANFO calculated from the consumption report of the mining sites is 2% [60]. The N30 is used as an initiator for the ANFO. The Company SOTEMU produces both explosives in Tunisia. The ANFO explosive is a granulated product whose commercial name is Sismex [186]. It is composed of 94% of Ammonium Nitrate and 6% fuel. The N30 Nitrate explosive is a powdery nitrate explosive encapsulated in a plastic cartridge. The commercial name is Nitrex. These are explosives whose essential constituent is 80% of Ammonium Nitrate, combined with 20% of a pure explosive Pentrite, according to the product description of the company SOTEMU [186].

In Ecoinvent, the process of producing explosives initially integrates the work of Kellenberger (2007) for the life cycle inventories of building products [119]. The explosive used is a Tovex product, a slurry explosive produced in Switzerland differs in composition and characteristics from the ANFO explosives Sismex (powder explosive). As no information on the production process in the SOTEMU company was available, the process "Explosive - Tovex" was used for the production of nitrate explosives. The production, the transport to the mining site, and the blasting operation are all integrated into the process: "Blasting phosphate-TN". The data about blasting was calculated according to the CPG process [16]. Data about the blasting behavior of the explosive were taken from the French Professional Agency for Risk Prevention in Building and Civil Engineering (OPPBTP) [154]. The input-output table of the process "Blasting phosphate-TN" can be found in Annex D, Figure D-4.

Table 6.3: Flow inputs for the process "Mining operations phosphate rock, as P_2O_5-TN"

Flow name, Location	Unit	Per kg CPRT[a]	Per kg P_2O_5	Comment
Building, Hall, steel construction, RoW	m^2	4.09E-06	1.70E-05	Building for maintenance, Storage for explosives, Fuel, and others
Conveyor belt, GLO	m	1.03E-07	4.29E-07	Conveyor belt (L: 9km, 25yrs lifetime; Capacity: 1000 t/h)
Diesel, burned in agricultural machinery, GLO	MJ	1.74E-01	7.25E-01	Modified process Diesel the infrastructure is deleted only the fuel used in machinery
Electricity, medium voltage, TN	MJ	1.65E-04	2.48E-03	Electricity Industrial, Medium voltage
Fluorine, 4.5% in apatite, 2.4% in crude ore	kg	2.00E-02	1.93E-01	Elementary flow, resource in the ground
Hydraulic digger, RoW	Item(s)	1.06E-10	4.42E-10	Hydraulic excavator (500 t, 20yrs lifetime)
Industrial machine, heavy, unspecified, RoW	kg	1.41E-06	5.89E-06	Blasthole Drills DM45 Atlas Copco (35-41 t, 20yrs lifetime)
Industrial machine, heavy, unspecified, RoW	kg	3.61E-05	1.50E-04	Dumper (170 t, 20yrs lifetime)
Industrial machine, heavy, unspecified, RoW	kg	9.33E-06	3.89E-05	Bulldozer (66 t, 20yrs lifetime)
Lorry, 40 metric tons, RoW	Item(s)	2.12E-10	8.84E-10	Truck (85 t, 15yrs lifetime) -> doubled up because of weight difference
Lubricating oil, GLO	kg	1.72E-04	7.17E-04	Lubricating oil for machinery
Maintenance, lorry 40 metric tons, RoW	Item(s)	6.72E-10	2.80E-09	Maintenance for Machinery (19 Machines, 20yrs lifetime)
Blasting phosphate, TN	kg	2.21E-03	9.2E-03	ANFO: Granulated product in a bag of 25 kg
Occupation, construction site,	m^2*a	1.11E-05	4.61E-05	Elementary flow, land
Occupation, mineral extraction site	m^2*a	2.44E-04	1.02E-03	Elementary flow, land
Occupation, industrial area	m^2*a	2.77E-06	1.15E-05	Elementary flow, land
Phosphorus, 18% in apatite, 11% in crude ore	kg	1.10E-01	4.58E-01	Elementary flow, resource in the ground
Transformation to the mineral extraction site	m^2	4.09E-02	1.70E-01	Elementary flow, land
Transport, freight, lorry >32 metric tons, EURO3, RoW	kg-km	3.61E-02	1.70E-01	Transport of crude PR (by truck)
Transport, freight, lorry >32 metric tons, EURO3, RoW	kg-km	5.1E-05	2.13E-04	Transport of equipment (500 t), road (500 km)

Transport, freight, lorry >32 metric tons, EURO3, Row	kg-km	8.48E-04	3.53E-03	Transport of intermediate layer (with dumper)
Transport, freight, sea, transoceanic ship, GLO	kg-km	5.1E-04	2.13E-03	Transport equipment, (500 t) freight ship (5000 km)

a: CPRT: Crude Phosphate Rock Treated mechanical grinding and classification at<40 mm

6.4.1.3 Outputs of the process mining operations phosphate rock, as P_2O_5-TN

For the inventory of mining operations, only the emissions to air were considered. The reason is that the open-pit mining techniques do not generate waste going into the water bodies. The backfill of the overburden and intermediate layers are not considered waste, as they are returned to the ground, which is their original source.

The quantity of contaminant transferred between air and soil, or air and water is not possible to estimate with the available data. The assessment method ReCiPe in LCA can, however, estimate such transfer between different environmental compartments, which is referred to as fate or pathway according to the environmental mechanism involved (transport, dispersion, and deposition in the environment) [86]. However, the limit of the ReCiPe method is that some environmental mechanisms are dependent on regional conditions and weather parameters (wind direction and speed, rainfall, temperature), relevant to estimate the fate of the emission and exposure to the contaminant [86].

For instance, the existence of potential transfer of radioactive elements between the air and soil due to phosphate rock mining was a subject of a study about the natural radioactivity of the Gafsa soil in 1991 [28]. The authors discuss the hypothesis of the transfer of radioactive elements between air charged with dust above the mining sites and the surrounding soil. Due to the arid and dry climate in the region, the dust emitted by the mining activities and containing polluting elements is carried out by the wind to the surrounding areas [28]. The results of their study show that the soil radioactivity in Ra and U increases as it gets closer to mine sites.

- **Toxic gases**

For the blasting operation, the emissions to air were assumed to be mainly Carbon Monoxide (CO) and Nitrogen Oxides (NO_x). The calculation is based on the data found in the literature on the behavior of different Nitrate explosives [21, 154].

Blasting generates a large quantity of fume and gases in the air. The average volume of 1 kg of gas emitted is 975 l/kg humid gas [21]. Half of this is water vapor, and the remaining

50% is composed of 20-30% Nitrogen (N_2) and 15-40% Carbon Dioxide (CO_2), 2-15% Carbon Monoxide (CO), and 0.05-5% Nitrogen Oxides NO_x [21]. The use of 1 kg of Nitrate explosive generates approximatively the emissions of 25-35l of a mixture of Carbon Monoxide CO and Nitrogen Oxides NO_x in the air. The volume of the Carbon Monoxide and Nitrogen oxides are estimated to be CO+5NO_x l/kg of explosive [154]. The composition of toxic gases depends on the chemical and physical characteristics of the ANFO explosive, such as the content of the fuel [171, 174]. Table 6.4 shows the direct emissions generated by the blasting operation.

- **Particulate matter**

In addition to the toxic gases, dust and particulate matter (PM) are released into the air during blasting. An average dust ratio of 2 g/m^3 of gas emitted can be retained [21].

In addition to the blasting operation, the crushing and sieving of phosphate rock contribute to dust and particulate matter emissions. Particulate matter emission also happens during road transportation between the mining sites, as the truck is the main transportation mode of crude phosphate rock and the barren rock, and the transportation routes are mainly unpaved roads.

According to Tartakovsky et al. (2016), the contribution of using the bulldozer and shovel for the removal of the overburden is negligible in weather conditions of the Khneifiss open-pit phosphate mine situated in a desert area in Syria [188]. As Gafsa climatic conditions are comparable to the phosphate open-pit mining in Syria described in the study, the emission factor of removing the overburden by bulldozer was estimated to be 698 g PM/metric ton [188]. The European Environment Agency (EEA) estimated the emission factor for the net mining operations (excluding the processing such as crushing and the transport) to be 102 g PM/metric ton [76]. For this inventory, the Particulate matter PM was estimated according to the method of the (EEA) [76]. The EEA method consists of modeling the individual mining process steps using regional data and the data about the studied mineral production sites. Dust emissions in quarries come from multiple points, and the emission sources are distributed within a vast area, which could have different wind and precipitation properties [76]. They consist of total suspended particles (TSP), some of which are in the PM10 fraction and, to a smaller extent, also in the Pm^2.5 fraction [76].

The typical points of emission are:

- Drilling and blasting
- Material processing
- Internal transport
- Material handling operation: loading and unloading
- Wind erosion from stockpiles

For the calculation, a spreadsheet model was used, which was developed and made available by the EEA to calculate particulate emissions at the country level for "Quarrying and mining of minerals, other than coal" [75]. The data used to produce the worksheet was based on the AP-42 (US EPA) methodology: US EPA, "AP-42 (1998) for Mineral Products Industry - Western Surface Coal Mining" [198]. For the regional data, the metrological data of the different cities nearby the mining area were used, which are: Gafsa south district, Metlaoui, Redeyef, Moulares, and M'dhila. The wind erosion from stockpiles was not included as an emission point due to missing data about the number of stockpiles and the average time of the stockpiles on the storage area.

The results of the calculation of particulate matter emission factors for the mining operations are shown in Annex D, Table D-4. Internal transport has a very high emission factor of dust since 96% of the internal transportation occur on an unpaved road.

To calculate the emissions per 1 kg of crude phosphate rock and per 1 kg of P_2O_5, the emission factors were multiplied by the total activity extraction of the crude phosphate rock, which was 6.13 MMT in 2017, and divided by the total output of crude phosphate rock sent to the washing plant after the first mechanical treatment (crushing sieving and screening). The total output of crude phosphate rock sent to the washing plants was 4.89 MMT. The PM emissions during blasting and drilling were calculated using the quantity of the overburden and intermediate layer. The phosphate layers are not subject to blasting. The quantity of overburden and the intermediate layers were 80.4 MMT in 2017. In the E-LCA software, the flow emission to air for PM was assigned to the low-density population as the mining sites are outside of the urban areas.

- **Radioactive elements**

Phosphate rock deposit contains radioactive elements (radionuclides). The radioactivity in the phosphates comes from the uranium-238 (238U) and the thorium-232 (232Th). Minerals that contain a naturally radioactive element, exploited for their non-radioactive properties, and where human activities could increase the exposure of people to ionizing radiation are known by "Naturally Occurring Radioactive Material"-(NORM) [10].

Phosphate dust particles generally have the same specific activity as the phosphate rock from which the dust originates [198]. Therefore, the emission of radioactive elements to the air was estimated by multiplying the specific activity of the phosphate rock with the total suspended particles (TSP).

For this inventory, the data on the specific radioactive activity of phosphate rock was taken from the doctoral work of Fattah (2005) at the research center of the CPG, analyzing samples from four deposits at the CPG [78] and from the master research work of Khelifi (2012), analyzing samples from the phosphate deposit of Om Lakhchab at the CPG [121]. The radionuclides which were subject to the mentioned studies are the Uranium-238 (238U), the Thorium-232 (232Th), the radium-226 (226Ra), and the Potassium-40 (40K). The average radioactivity of the Tunisian phosphate deposit, according to the data, is 400 Bq/kg 238U, 30 Bq/ 232Th, 325 Bq/kg 226Ra, and 61 Bq/kg 40K (see Table 6.5).

- **Cadmium**

Cadmium (Cd) is an element associated with the formation of sedimentary phosphates of marine origin. The levels of Cd in the Gafsa phosphate basin varies from deposit to another. In the same deposit, the content of Cd also varies from a layer to another. The average for the Gafsa phosphate basin is 36.7 ppm. It varies between 3.3 ppm in layer number 4 and 5 in the deposit of Om El Khecheb [121] and 86 ppm in layer number 6 in the deposit of Kef Schfayer, according to laboratory analysis conducted by the CPG [59]. For this inventory, the emission of cadmium to the air is calculated by multiplying the average content of crude phosphate rock, which is 36.7 ppm, with the emission factor of TSP for the mining operations (see Table 6.5), excluding the blasting and drilling, which are calculated separately to the flow process "Blasting phosphate-TN" (see Table 6.4).

Table 6.4: Outputs of the Blasting operation

Output	Unit	Per kg crude PR	Per kg P_2O_5
Carbon Monoxide CO	kg	1.07E-02	4.46E-02
Nitrogen Oxides NO_x	kg	5.73E-02	2.39E-01
Total particulate TSP	kg	2.55E-04	1.06E-03
PM < 10µm	kg	1.34E-04	5.58E-04
PM < 2.5 µm	kg	1.32E-04	5.49E-04
^{238}U	kbq	1.02E-04	4.26E-04
^{232}Th	kbq	7.67E-06	3.19E-05
^{226}Ra	kbq	8.29E-05	3.45E-04
^{40}K	kbq	3.95E-09	1.65E-08
Cd	kg	9.39E-09	3.91E-08

Table 6.5: Outputs of the process "Mining operations phosphate rock, as P_2O_5-TN."

Output	Unit	Per kg crude PR	Per kg P_2O_5
Total particulate TSP	kg	8.90E-03	3.71E-02
PM < 10µm	kg	2.56E-03	1.07E-02
PM < 2.5 µm	kg	2.59E-04	1.08E-03
^{238}U	kbq	3.57E-03	1.49E-02
^{232}Th	kbq	2.68E-04	1.12E-03
^{226}Ra	kbq	2.90E-03	1.21E-02
^{40}K	kbq	5.47E-04	2.28E-03
Cd	kg	3.28E-07	1.37E-06

6.4.2 Wet beneficiation phosphate rock, as P_2O_5-TN

6.4.2.1 Process description

After crushing and screening, the crude phosphate rock, which has a size of less than 40 mm and contains an average of 24% P_2O_5, will be mixed with water in the scrubber and undergoes operations of screening, classification, and filtration to produce a concentrate of phosphate rock containing 29% of P_2O_5. The wet beneficiation aims to separate impurities and coarse from phosphate rock. The steps of wet beneficiation at the CPG washing plants are: (1) Scrubbing and disintegrating, (2) Screening, (3) Classification, and (4) Filtration and concentration (see Figure 6.5).

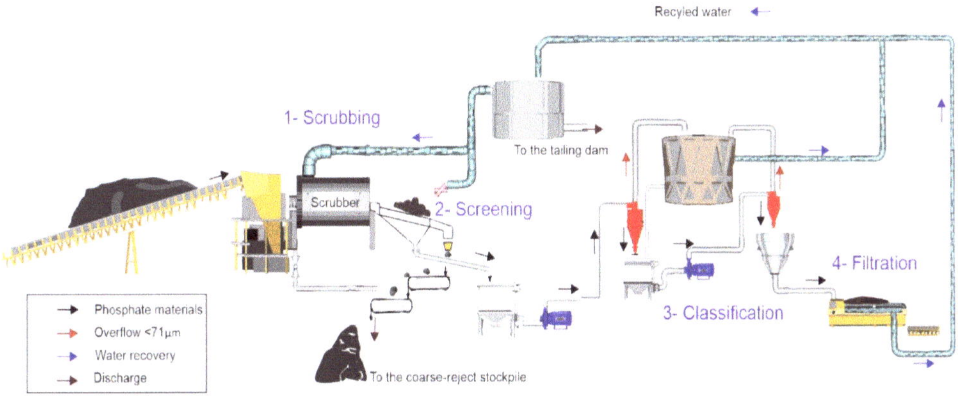

Figure 6.5: Simplified process flow of the beneficiation of phosphate rock based on the synoptic of the industrial process at CPG [167]

- **Scrubbing and disintegration**

The wet beneficiation process starts with feeding the crude phosphate rock into a rotating scrubber. The material is mixed with industrial water at 66% water and 33% phosphate rock [61]. Due to the rotational movement, the material inside the scrubber undergoes attrition and disintegration of hard rock. It transforms into a pulp.

- **Screening**

The pulp is poured on a vibrating screen of 2 mm mesh with jets of water under pressure for the removal of clay and sand and to separate the coarse rejects, which are larger than 2 mm. The coarse rejects are sent to the stockpile, containing 10-15% P_2O_5 [166] (see Sankey diagram, Figure 6.3). The coarse rejects represent 15% of the crude phosphate rock mass entering the plant [61] (see Figure 6.7).

- **Classification**

The pulp containing materials < 2 mm, is injected by pumps into hydro-cyclones to separate the particles at 71 µm. The hydro-cyclone separates the underflow fractions of 71 µm – 2 mm, which is phosphate particles concentrated at 29% of P_2O_5, and the overflow fraction, which contains mainly fine particles of clay under 71 µm. The overflow is injected in a decantation tank in the wastewater treatment plant for dewatering.

- **Filtration**

The 71 µm - 2 mm size phosphate particles are conveyed on a filter belt to reduce the water content of the product to 15% [61]. The product coming out on the filter belt is the beneficiated phosphate rock, which is the final marketable product to be sent to the chemical plant or the drying unit. The beneficiated phosphate rock is concentrated at 29% P_2O_5. The final product represents about 65% of the total mass at the washing plant entrance [61].

- **Flocculation and water recycling**

The process of wet beneficiation generates a large volume of slurry. The decantation step alone does not allow efficient dewatering and the settling of fine particles. Thus, it poses a problem not only to dump vast quantities of slurry in lagoons but also to lose a considerable volume of water, which is pumped from the underground with energy. Also, water resources in the Gafsa region are limited. Therefore, water recycling is an essential solution to reduce the consumption of underground water resources and to improve the disposal of slime by discharging a smaller quantity of solid waste rather than a large quantity of slurry.

As the slime is a colloidal suspension, the aggregation and separation of particles from water are possible only with the addition of flocculants. Flocculants are clarifying agents known as polyacrylamide, which are water-soluble polymers. They increase the viscosity of water and improve the aggregation of a particle suspended in the solution by retaining large amounts of water [157].

At the CPG water treatment plants, the flocculants product is mixed with clean water in a flocculation station (see Figure 6.6). The solution is concentrated at 4 g/l, and then the solution is diluted in water at a dilution ratio of 1:10 [61]. The flocculated water is pumped in the decantation tank where the clarification of water takes place. The clarified water is moved to the basin mixed with the newly drilled water. The concentrated slime (the range of wt. solid concentration is 126-200 g/l) is discharged into a dam outside the plant [61].

For this inventory, the flotation step to recover phosphate from slime was not considered as the flotation process was stopped in the last two years. The discharged slime contains around 12-15% of P_2O_5.

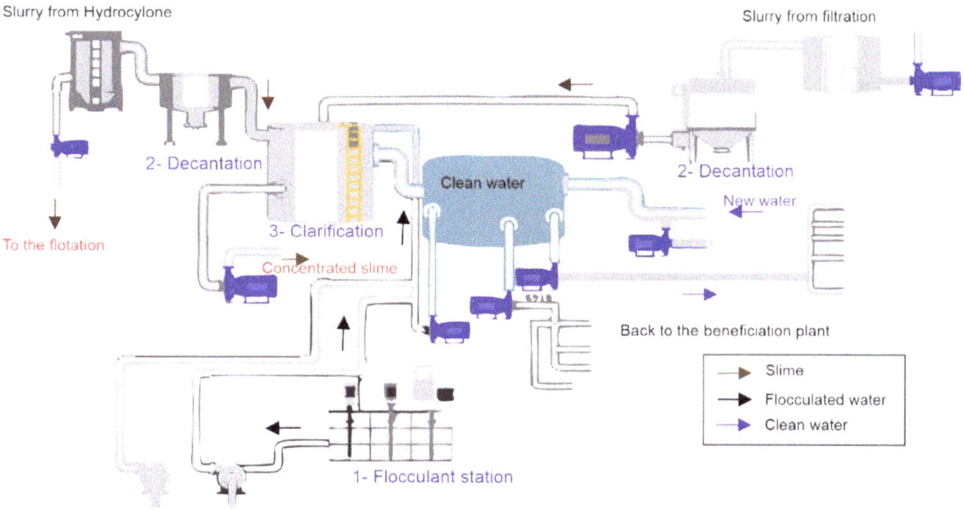

Figure 6.6: Simplified process of flocculation of slime and water recycling based on the synoptic of the industrial process at CPG [167]

6.4.2.2 Inputs of the process wet beneficiation phosphate rock, as P_2O_5-TN

The operative resources are the crude phosphate rock, water, the industrial plant to perform the scrubbing, classification, and filtration, and energy to run the beneficiation plant. Table 6.7 displays the inputs and their flow names per 1 kg beneficiated phosphate rock and per functional unit of 1 kg P_2O_5 for the inventory of “Wet beneficiation phosphate rock, as P_2O_5-TN”.

- **Energy**

In contrast to mining operations, the beneficiation process uses more electric energy than diesel. The scrubber, the vibrating screen, as well as the hydro-cyclone, and the filtration use electric energy and pneumatic energy, which is produced by the electric energy. For this inventory, the total electric energy consumption is collected from the reports of the washing plants at CPG. The total consumption of washing, screening, and flocculation is 0.041 MJ of electric energy per kg of beneficiated phosphate rock at 29% P_2O_5. The consumption of diesel is 0.0036 MJ per kg beneficiated phosphate rock at 29% P_2O_5. The final energy consumption of the beneficiation process is 0.045 MJ per kg beneficiated phosphate rock at 29% P_2O_5 and 0.15 MJ per kg of P_2O_5. The same as the mining operations, the electricity is supplied by the Tunisian company of electricity and gas (STEG). The input process flow is “Market of electricity medium voltage-TN.”

- **Water**

In the beneficiation plants, the average quantity of water used is 3.68 m^3 per metric tons of beneficiated phosphate rock. The optimum targeted by the company is 3 m^3 per metric ton of beneficiated phosphate rock [61]. However, because of the inefficiency of the hydro-cyclone separation and the water recycling processes, the water consumption is higher than the desired objective [61]. According to Figure 6.7, 55% of the water is reused in the washing process (262 t/h), which is equivalent to 2.03 m^3 of recycled water per metric ton of beneficiated phosphate rock and 45% of new industrial water (213 t/h) which is equivalent to 1.65 m^3 of industrial water per metric ton of beneficiated phosphate rock. As the water is drilled groundwater, the input process flow used for this inventory is "Water well in the ground."

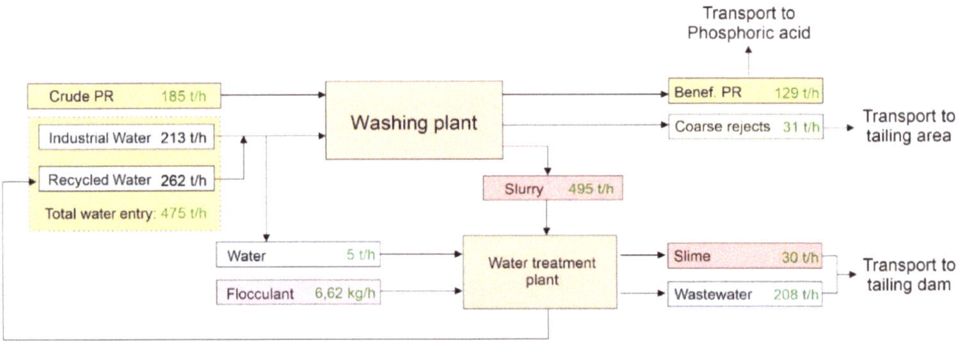

Figure 6.7: Material balance of the beneficiation process according to average values from the beneficiation plants at the CPG in 2017 [61]

The source of water used at the beneficiation plants is groundwater drilled in a deep borehole. The water is designated as "industrial water", as its salinity does not qualify the water to be used as drinking water or for irrigation in agriculture. The salinity of the drilled water is 6-12 g/l [165]. The number of groundwater drill boreholes is 12. Table 6.6 shows the names of the sources of the deep groundwater boreholes used by the different beneficiation plants. Water loss occurs during the transfer from the drilling station to the beneficiation plants. It is estimated to be 10-12% of the total annual groundwater drilled by the company [173]. For this inventory, 10% of industrial water was added on top of the annual industrial water consumption of the CPG.

Table 6.6: Water boreholes as sources of industrial water supply to the beneficiation plants [182]

Beneficiation plant	Water boreholes source
Metlaoui beneficiation four plants	Gouifla, Segui, Tarfaoui, Berka
Moulares beneficiation plant	Moulares
Redeyef beneficiation plant 1 and 2	Tarfaya
M'dhila beneficiation plant 1, 2 and 3	Gouifla, Segui, Tarfaoui

The washing of the pebbles of crude phosphate rock consumes a large quantity of industrial water. The annual industrial water drilled in 2017 is estimated to 12 Mm^3 (the annual water drilled can vary according to the total production of beneficiated phosphate rock and the efficiency of water recycling).

- **Flocculant**

The flocculants used in the recycling of water from the slurry of phosphate is "Slime floc 150", which is an anionic polyacrylamide. The flocculants are produced by the company SNF China. The product is a powder form. The flow attributed to the flocculants is "Market for polyacrylamide-GLO".

- **Transport**

Transport of the fine coarse rejects is assured by the trucks of the company STTPM. As the trucks used for the transport were already included in the mining operations, the transport module was customized to take the transport part into consideration and exclude the production of the truck itself.

- **Infrastructure**

The area occupied by the ten beneficiation plants is 1.06E+04 m^2, which represents 5.31E+02 m^2/ year (operation time estimated as 20 years). It is less than 0.05% of the entire occupied area of 1.20E+06 m^2 (see Table 6.2). For the industrial plants' equipment, only the scrubber and the hydro-cyclones were included in the inventory as information about their weight and the average period of use in the plant are available.

The average weight of the scrubber is 23 metric tons [181], and he has a lifetime of two years. The average weight of hydro-cyclone is 125 kg and has an estimated lifetime of twenty years [131]. In every plant, there is one scrubber and 6 to 7 hydro-cyclones. For the various industrial machines, the flow "Industrial machine, heavy unspecified" was attributed. The plant infrastructure was estimated according to the measure on the satellite picture (see Annex D, Figure D-2). The flow "Building hall, steel" was attributed to the beneficiation plants. The occupation of the industrial area was assumed to be 20 years, and the occupation of the construction was assumed to be 5 years.

Table 6.7: Flow inputs for the process "Wet beneficiation phosphate rock, as P_2O_5-TN."

Flow name, location	Unit	Per Kg benef. PR	Per Kg P_2O_5	Comment
Conveyor belt, GLO	m	3.26E-06	1.12E-05	Considered for the filtering band
Building hall steel construction, RoW	m^2	2.73E-06	9.43E-06	Beneficiation Plant building
Industrial machine, heavy unspecified, GLO		3.34E-05	1.15E-04	Industrial equipment: scrubber and hydro-cyclones
Wastewater treatment facility, capacity 4.7E10 L/year, GLO	item	2.89E-10	9.97E-10	Water treatment and recycling facility
Occupation construction site	m^2*a	6.143E-07	2.12E-06	Occupation construction of the industrial area for 5 years, Elementary flow
Occupation, industrial site	m^2*a	1.535E-07	5.30E-07	Occupation of the industrial site 20 years, Elementary flow
Electricity medium voltage, TN	kWh	1.14E-02	3.93E-02	
Diesel burned in building machine, GLO	MJ	3.65E-03	1.26E-02	Diesel for the machinery (45.6 MJ per kg diesel)
Lubricating oil, GLO	kg	1.80E-05	6.22E-05	
Water, well, in-ground,	m^3	3.4E-03	1.17E-02	Elementary flow
Polyacrylamide, GLO	kg	7.32E-05	2.52E-04	Flocculants Slimefloc 150
Transport, freight, lorry>32 tons, EURO3, RoW	kg-km	1.74	5.99	Excluding the diesel consumption, Transport of beneficiated rock lorry
Transport, freight, lorry>32 tons, EURO3, RoW	kg-km	3.4	11.7	Excluding the diesel consumption for the transport of fine coarse rejects
Conveyor belt production, GLO	m	1.43E-06	4.94E-06	Transport via a conveyor belt (25 yrs, lifetime; 1000 t/h and 3500h/year)

6.4.2.3 Outputs of the process wet beneficiation phosphate rock, as P_2O_5-TN

The emissions of the process "Wet beneficiation phosphate rock as P_2O_5-TN" are displayed in Table 6.8. The wet beneficiation process does not produce emissions to air, such as dust or radioactive elements, as the crude phosphate rock is mixed with water during the entire process. The CO_2, NO_x, and CO emissions due to the burning of fuel in the machinery are included in the linked processes. However, the production of pneumatic energy leads to the generation of waste heat (thermal energy is the waste of the air compression process). The amount of waste heat was not possible to calculate or estimate from the available data.

- **Emissions to water**

The emissions to water are mainly sulfate, phosphorus, and fluorine due to the leaching[8] of those elements from the solid fraction of the slime into the water fraction. The routes of contamination of water by phosphorus, sulfate, and fluorine are mainly (1) the leakage of the canal during the transfer of the slurry from the beneficiation plant to the settling ponds, (2) the leakage of the settling ponds, and (3) the contamination of municipal canalization by the beneficiated phosphate rock lost from the trucks during the road transport. In 2016, 1,205 metric tons of beneficiated phosphate rock were lost during road transport, and 6 metric tons were lost in 2017 [53, 55].

The slurry is carried out of the washing plants under gravitational force into an excavated canal of 130 m, located 15-20 km outside of the city [173]. However, the hydrographic network such as rivers and aquifers were reported to be contaminated by inorganic pollutants and toxic metals due to the beneficiation of phosphate rock [18, 94, 120, 138, 173, 182]. The contamination of the river-beds is evident for the visitor of the mining basin of Gafsa. The water pollution can be visualized on google maps (see Annex D, Figure D-4).

The data on the chemical analysis of the slurry were taken from the work of Smida (2012). The analysis of slurry discharged from the beneficiation plants of Redeyef and M'dhila was conducted by collecting samples from different points along the slurry transfer canal (from the exit of the plant to the settling pond) [182]. The study of the mobility of elements from the solid fraction (slime) to the liquid fraction (water) of the slurry shows that the mobility of sulfate, fluorine, and phosphorus is very high. However, the mobility of Cd, U, and Pb is very low [182]. For this inventory, the concentration of Phosphorus, Sulfate, and Fluorine were multiplied by the volume of the liquid fraction of the slurry designated as wastewater in Figure 6.7. The concentrations are displayed in Annex D, Table D-5.

[8] Leaching is a natural process by which water soluble substances are washed out from soil or wastes. These leached out chemicals (called leachates) cause pollution of surface and sub-surface water http://www.businessdictionary.com/definition/leaching.html.

- **Emissions to soil**

Soil contamination comes mostly from the infiltration and accumulation of heavy metals from the settling ponds to the surrounding soil. Heavy metals such as Cd, U, and Pb have low mobility from the solid fraction to the liquid fraction of the slurry (due to leaching). They accumulate in the soil [182]. The data for soil contamination were taken from Smida (2012). The concentration of heavy metal was multiplied by the amount of the solid fraction of the slurry (see Figure 6.7). The concentrations of heavy metals in the solid fraction of the slurry (slime) are displayed in Annex D, Table D-6.

Table 6.8: Outputs of the process "Wet beneficiation phosphate rock, as P_2O_5-TN"

Output	Unit	Per kg benef. PR	Per kg P_2O_5
Emission to water			
Fluorine (F)	mg/kg	89.465	308.500
Sulfate (SO_4^{2-})	mg/kg	4892.034	16869.082
Phosphate ion (PO_4)	mg/kg	47.207	162.783
Nitrate (NO_3^-)	mg/kg	33.692	116.180
Nitrite (NO_2^-)	mg/kg	29.123	100.426
Cadmium Cd	mg/kg	0.0190	0.0656
Lead (Pb)	mg/kg	0.0019	0.0065
Zinc (Zn)	mg/kg	0.100	0.347
Emission to soil			
Cadmium (Cd)	mg/kg	7.125	24.571
Lead (Pb)	mg/kg	1.202	4.147
Uranium (U)	mg/kg	8.300	28.623
Zin (Zn)	mg/kg	74.752	257.766

6.4.3 Drying of beneficiated phosphate rock

The operation of drying concerns only the wet beneficiated phosphate rock, which will be exported. The drying operation aims to facilitate the handling of materials at the port and loading to the ship.

The wet beneficiated phosphate rock has an average moisture content of 10-15%. However, it has to be dried to 1-3% moisture content when exported [66].

The production of dry, beneficiated phosphate rock is marginal. The annual production of dry beneficiated phosphate rock is around 2% of the total annual production at the CPG [53, 55]. The CPG sells the wet beneficiated phosphate rock on the local market to the phosphoric acid plants (GCT). However, the company kept the drying operation to meet the occasional export market demand [46].

6.4.3.1 Process description

The wet beneficiated phosphate rock is dried in a direct-fired rotary drier at 120°C. A simplified industrial process is shown in Figure 6.8. The wet beneficiated phosphate rock is transferred by a conveyor to be poured into a hopper. The material is tumbled into a rotary drum where it meets the hot air coming from the boiler room placed in the entrance of the rotary drum. The beneficiated phosphate rock advances while being stirred by the rotation of the drum and the bucket inside the drum.

This operation helps to increase the contact surface between the product and the hot gas. The dry phosphate falls on a conveyor, which carries it to the stockpile area. An exhaust ventilation system blows the dust to a cyclone collector and then to the dust collecting chamber. The dust which settles on the bottom of the chamber consists of the fine particles of phosphate rock. The gases are loaded with reduced dust, and the vapor is emitted into the air through a smokestack.

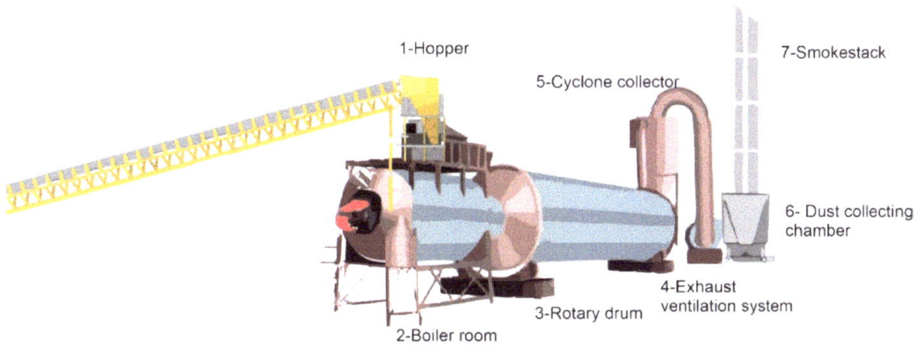

Figure 6.8: Simplified process of rotary drying of beneficiated phosphate rock at the CPG

6.4.3.2 Inputs of the drying process

- **Infrastructure**

There are three active drying units at the CPG. The drying units are placed on the same sites as the beneficiation plants. Therefore, the land occupation flow process was not included to avoid double counting. For the plant itself, the flow "Heavy industrial machine, unspecified" was attributed to the rotary drum. The rotary drum dimensions are 25 m in length and 2.5 m in diameter. The weight of the rotary drum was estimated based on the dimensions of the dryer from the "911 Metallurgists & Mineral Processing Engineers" website [132].

There are three active drying units at the CPG. The drying units are placed on the same sites as the beneficiation plants. Therefore, the land occupation flow process was not included to avoid double counting. For the plant itself, the flow "Heavy industrial machine, unspecified" was attributed to the rotary drum. The rotary drum dimensions are 25 m in length and 2.5 m in diameter. The weight of the rotary drum was estimated based on the dimensions of the dryer from the "911 Metallurgists & Mineral Processing Engineers" website [132].

- **Energy**

Currently, the drying process uses industrial heavy fuel oil n°2 (HFO) as an energy source for heating. The average consumption of HFO at the CPG between 2016 and 2017 is 17 kg/metric ton of dry beneficiated phosphate rock. The consumption varies according to the amount of initial moisture of the wet beneficiated phosphate rock entering the dryer. Currently, the average moisture content of the wet beneficiated phosphate rock is 15%. In order to reduce heavy fuel oil consumption to 10 kg per metric ton, the initial moisture content of the beneficiated phosphate must be reduced to 10-11%.

6.4.3.3 Outputs of the drying process

The emissions of the process "Drying phosphate rock, as P_2O_5-TN" are mainly emissions into the air as dust, Fluorine, and water vapor. The input-output breakdown is displayed in Table 6.9.

- **The total particulate matter (TPM)**

The emission of PM was calculated according to the data on emission factors of the phosphate drying process of the EPA "Report AP-42 Chapter: 11 SECTION 11.21" [65, 66]. Emissions from dryers depend on the moisture content, the type of rock (a higher clay content induces higher emission rates), and the technology for dust emission control [65]. The dust emissions from the dryer were controlled with a cyclone collector at CPG (see Annex D, Figure D-5). They are most commonly used for phosphate rock drying, according to the Ecoinvent report n°8 [65, 159]. The data used for this inventory were given in table 4-3 Ref. No.2 (no- scrubber) in the background report of *Report AP-42 chapter 11: SECTION 11.21* [65]. The EPA emission factor is 4.0E-04 kg/kg dry beneficiated phosphate rock [65].

However, to avoid double counting of the PM emitted during the fuel burning included in the furnace process, 1.2E-05 kg PM/MJ heat was subtracted from the emission factor of the EPA as suggested by the Ecoinvent report n°8 [159]. The subtracted value was calculated from the emission factor of 100% fuel oil n°2 with 1.73E-05 PM/MJ [159]. The calculation procedure of the PM emission is in Annex D, Table D-7.

The particle size distribution was assessed according to the Ecoinvent report n°8, based on the EPA's work for controlled particle emissions of drying [65, 66, 159]. The shares used for the size distribution are 3% for PM ≥ 10 µm, 8% for 2.5 < PM < 10 µm. and 89% for PM < 2.5 µm [159]. The flow emission to air for PM was assigned to a high-density population. The reason is that the dryer units are in urban areas.

- **Other potential emissions**

Other gaseous emissions, such as Fluorine and Sulfur Dioxide, are considered in this inventory. The emission factor of Fluorine is 0.0038 kg/metric ton of dry beneficiated phosphate rock, according to the EPA (1993) table 4-3 Ref. No.2 (no- scrubber) in the background report of *Report AP-42 chapter 11: SECTION 11.21* [65]. The average emission of Fluorine is 724 kg/year.

Emissions of Sulfur Dioxide (SO_2) due to the burning of heavy fuel oil are included in the linked process "Fuel burned in a furnace". Moreover, according to the EPA, when high-sulfur residual fuel oil, such as industrial HFO, is used, phosphate rock contains about 55% of lime (CaO), which reacts with Sulfur Dioxides (SO_2) to form Calcium Sulfites and Sulfates and thus reduces Sulfur Dioxide emissions [66].

The emission of water vapor was calculated based on the moisture content difference of the wet beneficiated phosphate rock at the entry of the dryer and the moisture content of the dry beneficiated phosphate rock at the exit of the dryer. It is assumed that the emitted water vapor in the air is 160 kg/ metric ton dry beneficiated phosphate rock.

Table 6.9: Input-Output of the process "Drying phosphate rock, as P_2O_5-TN."

Flow name, Location		Per kg dry benef. PR	Per kg P_2O_5	Comment
		Input		
Industrial machine, heavy unspecified, GLO	kg	2.36E-04	8.15E-04	
Heavy fuel oil, burned in refinery furnace, RoW	MJ	6.92E-01	2.39E+00	
		Output		
TPM	kg	3.88E-04	1.31E-03	Total PM 100%
PM≥10µm	kg	1.16E-05	3.92E-05	3% TPM
10µm >PM> 2.5µm	kg	3.10E-05	1.05E-04	8% TPM
PM<2.5 µm	kg	3.45E-04	1.16E-03	89% TPM
Fluorine (F)	kg	3.80E-06	1.28E-05	
Water vapor	kg	1.61E-01	5.54E-01	

6.5 Life cycle impact assessment (LCIA)

In this section, the environmental midpoint impact categories are discussed according to environmental areas of damage: water pollution, water scarcity, climate change, human health damage due to hazardous chemicals emission, soil pollution, and arable land use (see Figure 6.2, section 6.3.4).

The open-source software openLCA version 10.1 was used for the modeling of the LCIA [88]. The model allows assessing the environmental impact of each process, from mining operations to the drying of the beneficiated phosphate rock. The cut-off criteria applied to the calculation is 1%.

Drying of phosphate rock will be discussed separately (section 6.6) as its yearly production is minimal (2% of total production) comparing to the wet beneficiated phosphate rock.

6.5.1 General results

The complete results of the impact assessment of assessed processes. "Mining operations phosphate rock, as P_2O_5-TN", the "Wet beneficiation phosphate rock, as P_2O_5-TN" and the "Drying phosphate rock, as P_2O_5-TN" are displayed in Table 6.10.

6.5.2 Water depletion

The total amount of water depletion caused by the product system "Wet beneficiation phosphate rock, as P_2O_5-TN" is 0.96 m^3/ kg of P_2O_5. The direct contribution of the process is 1.2% of the total amount (0.012 m^3/kg P_2O_5). The main contributors are background processes belonging to the processes "Market of wastewater treatment facility" (59.54%), the process "Mining operations phosphate rock, as P_2O_5-TN" (21%), and the process "Market of conveyor belt" (14.8%). These input unit processes have no direct contribution. However, their background processes are responsible for 96% of the total amount of water depletion (see Annex D, Table D-9).

Table 6.10: Impact analysis results of the process "Mining operations phosphate rock, as P_2O_5-TN", "Wet beneficiation phosphate rock, as P_2O_5-TN" and "Drying phosphate rock, as P_2O_5-TN". ReCiPe Midpoint (H)

Impact category	Unit	Mining operations	Wet beneficiation	Drying
Water depletion	m^3	2.51E-01	9.63E-01	1.13E+00
Climate change	kg $CO_{2\ eq}$	1.81E-01	3.75E-01	5.92E-01
Photochemical oxidantfor-mation	kg NMVOCs	2.42E-01	2.43E-01	2.87E-01
Human toxicity	kg 1,4DB $_{eq}$	1.08E-01	2.47E-01	3.01E-01
Terrestrial acidification	kg $SO_{2\ eq}$	1.36E-01	1.37E-01	1.62E-01
PM10	kg PM10 $_{eq}$	7.12E-02	7.18E-02	8.02E-02
Ionizing radiation	kg ^{235}U	1.75E-02	2.49E-02	3.83E-02
Agri land occupation	m^2/year	2.75E-03	6.95E-03	7.82E-03
Freshwater ecotoxicity	kg 1,4DB $_{eq}$	9.40E-04	5.22E-03	6.19E-03
Terrestrial ecotoxicity	kg 1,4DB $_{eq}$	6.28E-05	1.84E-03	2.20E-03
Freshwater eutrophication	kg P_{eq}	1.97E-05	1.50E-04	1.80E-04
Natural land transformation	m^2	5.57E-05	7.62E-05	1.40E-04

6.5.3 Freshwater pollution

The pollution of freshwater is expressed by two midpoint impact categories; eutrophication and ecotoxicity.

- **Freshwater Eutrophication**

Eutrophication is the enrichment of the aquatic environment in nutrients (Phosphorus and Nitrogen), causing algal blooms, which results in disrupting the aquatic ecosystem balance [86].

The eutrophication is expressed by the amount of phosphorus equivalents emission in water (kg Peq). The production of wet beneficiated phosphate rock generates a total of 1.5E-04 kg Peq per kg P_2O_5. The main contributor is the wet beneficiation process itself. The direct contribution is 36.23% of the total eutrophication potential. The remaining contributors are background processes located in the unit process "Market of wastewater treatment facility" (40%), the process "Mining operations phosphate rock, as P_2O_5-TN" (11.17%), and the process "Market of conveyor belt" (10.82%). These unit processes have no direct contribution (see Annex D, Table D-10).

The high eutrophication potential is explained by the water contamination by the slurry conveyed out of the washing plants and the contamination by the loss of beneficiated phosphate rock during the road transport to the phosphoric acid plants (see section 6.4.2.3). The wet beneficiation process is responsible for the emission of 162.78 mg Phosphate/kg P_2O_5, 116.18 mg Nitrate/kg P_2O_5, and 100.42 mg Nitrite/kg P_2O_5 (see Table 6.8).

- **Freshwater Ecotoxicity**

Freshwater ecotoxicity refers to the potential of pollutants emitted to air, water, and soil, to have a toxic effect on freshwater ecosystems. Ecotoxicity, as calculated in the ReCiPe method, describes the environmental persistence, exposure, and the toxic effect of pollutants, such as heavy metals, on the ecosystem. The characterization factors of different pollutants are expressed using the reference unit of kg 1,4-Dichlorobenzene equivalents (kg 1,4 DB eq) [1].

The production of wet beneficiated phosphate rock generates 5.22E-03 (1,4 DB eq) per kg P_2O_5. The direct contribution is 1.85% of the total ecotoxicity potential. The main contributors are background processes located in the unit process "Market of wastewater treatment facility" (64.57%), the process "Market of conveyor belt" (16.12%), and the process “Mining operations phosphate rock, as P_2O_5-TN” (14.1%) (see Annex D, Table D-11). These unit processes have no direct contribution. However, their background processes are responsible for 94.8% of the total amount of ecotoxicity potential.

6.5.4 Soil pollution

Soil pollution is defined by two midpoint impact categories; terrestrial acidification and terrestrial ecotoxicity.

- **Terrestrial acidification**

Soil acidification is the phenomenon caused by acid deposition due to “acid rain”. Acidic gases emitted in the air, such as Sulfur Oxides SO_x, Nitrogen Oxides NO_x, and Nitrogen NH3, react with water in the atmosphere to form “acid rain” [1]. The acidification of soil causes damage to plants and organisms in the soil (such as microorganisms essential to maintain good soil health). It leads to a change in the terrestrial ecosystem. Acidification potential is expressed using the reference unit kg SO_2 eq.

The total acidification potential of the production of wet beneficiated phosphate rock is 0.137 kg SO_2 per kg P_2O_5. The beneficiation operation has no direct impact. However, the unit process “Mining operations phosphate rock, as P_2O_5-TN” is the direct and main contributor to the terrestrial acidification. It contributes 98.8% of total terrestrial acidification potential (see Annex D, Table D-12).

One of the mining operations, which is responsible for the emission of acidic gases, is blasting. During the blasting of the overburden and the intermediate layers of phosphate rock, the explosion produces a large volume of gases, mainly Carbon Monoxide CO and Nitrogen Oxides NO_x. The mining of 1 kg of crude phosphate rock generates an average of 0.0573 kg of NO_x (equivalent to 0.24 kg of NO_x per kg P_2O_5) (see Table 6.4)

- **Terrestrial ecotoxicity**

Terrestrial ecotoxicity refers to the potential of pollutants such as heavy metals – emitted to air, water, and soil – to have a toxic effect on soil organisms, which are crucial for soil functioning and maintaining the balance of its ecosystem. In the same way as freshwater ecotoxicity, it is calculated in the ReCiPe method based on environmental persistence, the exposure, and the toxic effect of pollutants. The terrestrial ecotoxicity is expressed in kg 1,4-Dichlorobenzene equivalents (Kg 1,4 DB eq) [1].

The total terrestrial ecotoxicity of the production of beneficiated phosphate rock is 1.84E-3 kg 1,4 DB eq per kg of P_2O_5. The beneficiation operation is the direct and foremost responsible for the emission of soil pollutants with 96.9% of total terrestrial ecotoxicity potential (see Annex D, Table D-13).

For instance, the outputs of the beneficiation operation are heavy metals, Fluorine, Sulfate, and Nitrate (see Table 6.8). One of the main soil contaminants is Cadmium Cd (7.12 mg/kg P_2O_5). The surrounding lands are contaminated due to heavy metals' infiltration from settling ponds, where the slurry is conveyed out of the washing plants.

6.5.5 Arable land use

The damage to ecosystems due to the use and the transformation of arable land into a mining area is assessed by two midpoint impact categories; the agricultural land occupation and the land transformation. The two impact categories are combined as occupation follows a transformation of the land [86]. For the mining operations, every year, a surface of 447,000 m^2 is transformed into the mining area. The average occupation period of a mining site was assumed to be 20 years. After the closure of the mining site, the site is left without rehabilitation.

Agricultural land occupation refers to the surface of arable land occupied by an industrial activity during a period of time. It is expressed in m^2 per year. The mining operations occupy the most extensive area as the beneficiation plants do not contribute directly to the use of arable land and its transformation. It was reported early in this chapter (see section 6.4.2.2) that the land occupied by the mining sites is transformed from desert to a mineral area. The total agricultural land occupation is 6.95E-03 m^2 per year/kg P_2O_5, and the total land transformation is 7.62E-05 m^2 per year/kg P_2O_5 (see Annex D, Table D-14).

The background process "Explosive production" is the main contributor to land occupation and land transformation (33.7% and 9.84%. respectively). However, the process of explosive production, taken from Ecoinvent, refers to the production of Tovex in Switzerland [119]. The production of explosive Sismex and Nitrex, produced by SOTEMU, is located in Guettar, Tunisia. Thus, it has been assumed that the production conditions are different between Tovex in Switzerland and Sismex in Tunisia. The results of land occupation and transformation have high uncertainty and cannot be attributed to the production of wet beneficiated phosphate rock in Tunisia.

6.5.6 Climate change

Climate change refers to the emission of greenhouse gases (GHGs) and their global warming potential (GWP). The rise of global temperature leads to the disturbance of ecological ecosystems by causing, for example, flooding, the rise of the sea level, and the extinction of species. Climate change also has negative impacts on human welfare, such as the increase of famine and water scarcity, the spread of diseases, and poverty for vulnerable communities.

The total emissions of GHGs generated by the wet beneficiate phosphate rock production is 0.37 kg CO_2 eq/kg P_2O_5. The main unit process contributor is the process "Mining operations phosphate rock, as P_2O_5-TN" (41% of the total emissions) (see Annex D, Table D-15).

During mining operations, the extraction of phosphate, and the transport of the overburden, diesel is used in heavy engines (like in the Dumper). Besides, the on-site transport of crude phosphate rock between the mining sites and the washing plants by lorry consumes a considerable amount of diesel. The consumption of diesel for mining contributes 18.2% of the total CO_2 eq emissions. The total transport operations (transport of crude phosphate rock, the coarse, and the beneficiated phosphate rock) contribute 5% of the total CO_2 eq emissions. The washing operation uses more electricity than fuel. Electricity production is also a source of CO_2 emission and contributes 7% to the total CO_2 eq emissions.

6.5.7 Human exposition to hazardous elements

- **Ionizing radiation**

According to the World Health Organization (WHO), ionizing radiation is a type of energy released by atoms and travels in the form of electromagnetic waves (gamma or X-rays) or particles (α, β, neutrons) [226]. The decay of radionuclide releases ionizing radiation. Radionuclides are unstable elements such as Uranium U, Thorium Th, and Lead Pb. The exposure of humans to high doses of radiation can be hazardous. Long-term exposure to low doses of radiation can also increase the risk of cancer and other long-term effects such as congenital disabilities. In the ReCiPe method, the midpoint characterization factors are taken at the fate and exposure level. The fate level is the radiation level calculated according to the type of radionuclide and the behavior of the radiation in the environmental compartment (air, water, soil). The exposure level is the dose that a human absorbs based on human body equivalence factors for the different ionizing radiation types (α-, β-, γ-radiation, neutrons) [86]. The impact category is expressed in a reference unit of 1 kg of uranium-235 equivalents (235U eq) to air. The total radiation potential of the wet beneficiated phosphate rock is 0.02 kg 235U eq/kg P_2O_5. The process "Mining operations phosphate rock, as P_2O_5-TN" is the main contributor to the ionizing radiation potential (64% of total ionizing radiation) (see Annex D, Table D-16). The radionuclides are mainly released in the air during the blasting, extraction, and loading and unloading of the trucks at the mining site. Those operations are responsible for 30.5% of the total ionizing radiation potential. Furthermore, the production of diesel and the explosive are also sources of ionizing radiation.

Although the wet beneficiation process generates waste containing uranium and contaminating in soil, the direct contribution appears to be null. The explanation is that the ReCiPe does not include a characterization factor for the radiation level in the compartment soil. The characterization factors for Uranium are only calculated for water and air compartment.

- **Particulate matter formation (PM10)**

Particulate Matter with a diameter of less than 10 µm (PM10) is a complex mixture of dust, soil, and chemicals emitted in the air. The long term exposure to PM10 causes health problems, especially respiratory and cardiovascular diseases, when inhaled [1]. The World health organization (WHO), classified the particulate matter as a cause of lung cancer [225]. According to the same source, PM10 is also the most widely used indicator to assess the health effects of exposure to ambient air pollution [225]. In the ReCiPe method, the midpoint characterization factor is the particulate matter formatting potential (PMFP) of a given pollutant. The PMFP is calculated as the intake fraction of PM10 from the given pollutant. The particulate matter formatting potential is expressed as kg PM10 equivalents (PM10 eq).

The total potential of particulate matter formation of the wet beneficiated phosphate rock production is 0.0718 kg PM10 eq/kg P_2O_5. The process "Mining operations phosphate rock, as P_2O_5-TN" is responsible for 99.22% of the total PM10 formation potential (see Annex D, Table D-17). Blasting, drilling, material handling, and transport using heavy trucks are responsible for the emission of small particulate in the air. Transport has the highest PM emission factor (see Annex D, Table D-4).

- **Photochemical oxidant formation (photochemical Ozone formation)**

Ozone is formed as the result of the reaction of non-methane volatile organic compounds (NMVOCs) and Nitrogen Oxides NO_x under the influence of sunlight [1]. Ozone has a hazardous effect on human health, especially on the respiratory system. Long exposure and high activity level during the exposure can lead to the irritation of the throat and airways [86].

In the ReCiPe method, the midpoint characterization factor is the photochemical ozone formation potential (POFP) of pollutants such as Sulphur Dioxide SO_2, Nitrogen Oxides NO_x, and NMVOCs [86]. The reference unit of POFP is kg NMVOC eq.

The total potential of Ozone formation of the wet beneficiated phosphate rock production is 0.24 kg NMVOC eq/kg P_2O_5. The process "Mining operations phosphate rock, as P_2O_5-TN" contributes to 99.7% of the total potential of ozone formation (see Annex D, Table D-18). The high ozone formation potential is caused by the emission of SO_2, due to the high diesel consumption, as well as the emission of the high amount of Carbon Monoxide CO and Nitrogen Oxides NO_x during blasting.

- **Human toxicity**

Human toxicity is defined as the potential of a chemical to affect human health. The human toxicity of a substance depends on its toxicity level, the dose of exposure, and the route of exposure such as inhalation, ingestion, or skin contact. Like the ecotoxicity of freshwater and soil, the human toxicity is calculated in the ReCiPe method based on the amount and concentration in an environmental compartment (air, water, and soil), the exposure route (inhalation, ingestion, and contact), and the toxicity level of pollutants. The human toxicity is expressed using the reference unit, kg 1,4-Dichlorobenzene equivalents (kg 1,4 DB eq) [1].

The total human toxicity potential generated by the production of the wet beneficiated phosphate rock is 0.247 kg 1,4 DB eq/kg P_2O_5. The main unit process contributors are the market of wastewater facility (41%) and the mining operations (40%). The beneficiation operation also contributes 6.6% of the total human toxicity potential. The main unit process contributors are the "Treatment of sulfidic tailing" (28.47%), which is a background process, located in the chain of the process "Market of wastewater treatment facility" and the process "Mining operations phosphate rock, as P_2O_5-TN" as a unit process (26.75%). The route of chemical emissions from mining operations is mainly emission into the air. The main chemicals emitted are Carbon Monoxide CO and Nitrogen Oxides NO_x

(due to Nitrate explosive), cadmium, radioactive elements, and PM10. According to the ReCiPe spreadsheet, the characterization factor of cadmium for human toxicity in the compartment air at a low-density population is 4.5E+04 kg 1,4 DB. The emission of cadmium to air is 1.37E-06 kg 1,4 DB /kg P_2O_5, equivalent to 6.2E-02 kg 1,4 DB/kg P_2O_5. That is precisely the amount of Human toxicity potential of the module "Mining operations phosphate rock, as P_2O_5-TN" (see Annex D, Table D-19). Therefore, cadmium emission has high human toxicity potential in the region.

6.6 Environmental impacts related to the process "Drying phosphate rock, as P_2O_5-TN"

Drying of wet beneficiated phosphate rock increases the total amount of environmental impacts since the output product of the process "Wet beneficiation phosphate rock, as P_2O_5-TN" is the input product of the process "Drying phosphate rock, as P_2O_5-TN". It needs 1.18 kg of P_2O_5 input from the process "Wet beneficiation phosphate rock, as P_2O_5-TN" to produce 1 kg of P_2O_5 from the process "Drying phosphate rock, as P_2O_5-TN". Drying also increases the content of the final product in P_2O_5. One kg of dry beneficiated phosphate rock has an average of 30% P_2O_5.

The drying process emissions to air are water vapor and particulate matter (see Table 6.9). The contribution to the potential of particulate matter formation PM10 is minimal (0.2% of total PM10 formation potential) (see Table 6.11).

The use of heavy fuel oil (HFO) also contributes to the increase in environmental impacts. Figure 26 shows that drying increases 81% of the climate change potential, 90% the radiation potential, and 166% the transformation of natural land. The main process contributor in the impact categories is the "Heavy fuel oil, burned in refinery furnace, RoW". The use of HFO contributes 67% of the increase of climate change potential, 61% of the increase of the radiation potential, and 71% of the increase in natural land transformation. The drying process increases the rest of the environmental impacts, with an average of 22%.

Table 6.11: Environmental impacts at the midpoint of the process "Drying Phosphate rock, as P_2O_5-TN."

Impact category	Unit	Total amount	Direct contribution	Main background contributor: Heavy fuel oil burning process
Climate change	kg $CO_{2\,eq}$	5.97E-01		1.78E-01
Human toxicity	kg 1,4DB $_{eq}$	3.00E-01		2.00E-02
PM10	kg PM10 $_{eq}$	7.90E-02	1.57E-04	5.30E-04
Ozone formation	kg NMVOCs	2.87E-01		5.60E-04
Ionizing radiation	kg ^{235}U	3.80E-02		1.10E-02
Water depletion	m^3	1.13E+00		4.50E-02
Freshwater ecotoxicity	kg 1,4DB $_{eq}$	6.19E-03		1.80E-04
Freshwater eutrophication	kg P $_{eq}$	1.80E-04		3.13E-06
Terrestrial acidification	kg $SO_{2\,eq}$	1.62E-01		2.00E-03
Terrestrical ecotoxicity	kg 1,4DB $_{eq}$	2.20E-03		2.65E-05
Agri land occupation	m^2/year	7.83E-03		2.30E-04
Natural land transformation	m^2	1.50E-04		6.61E-05

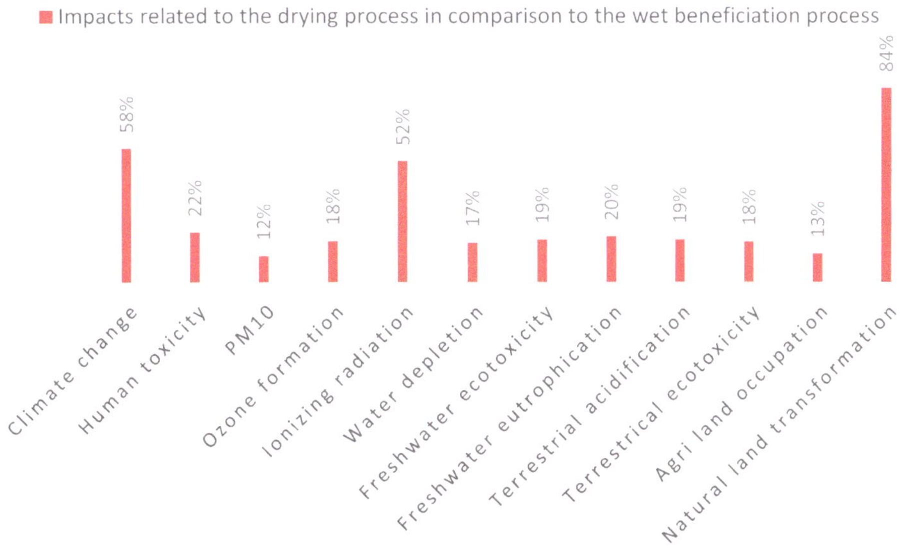

Figure 6.9: Environmental impacts related to the process "Drying phosphate rock, as P_2O_5-TN" in comparison to the process "Wet beneficiation phosphate rock, as P_2O_5-TN"

6.7 Interpretation

6.7.1 Main impact categories

The environmental hotspots located in the Gafsa region and related to the direct impact of mining operations and the beneficiation process are mainly the impact categories of the damaged area human health, such as Photochemical oxidant formation (242,749.74 metric tons of NMVOCs per year[9]). Climate change (134,415 metric tons of CO_2 eq per year), Human Toxicity (87,209 metric tons of 1,4 DB eq per year), PM10 emissions (65,201 metric tons of PM10 per year), and ionizing radiation (6,279 metric tons of 235U per year) (see Figure 6.10).

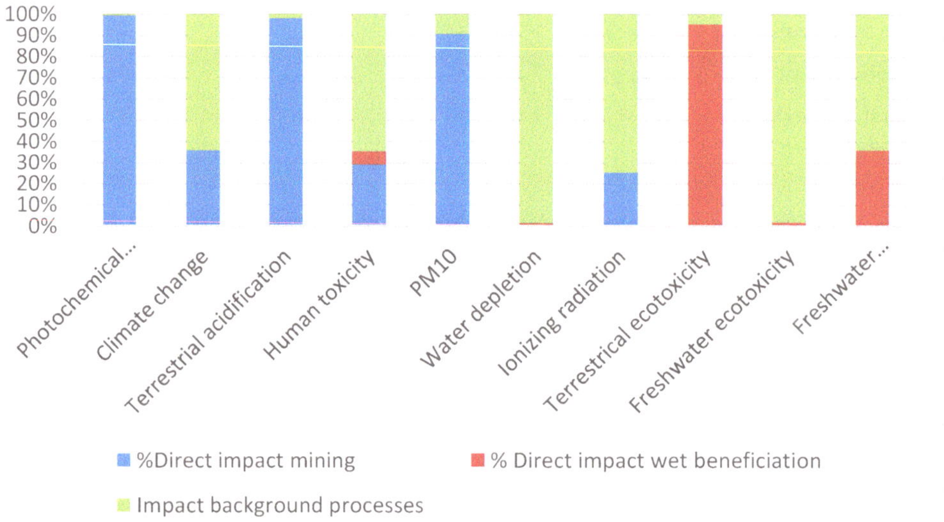

Figure 6.10: Main impact categories located in Gafsa due to direct emissions of the product system "Wet beneficiation phosphate rock, as P_2O_5-TN". Expressed per 1 kg P_2O_5

Moreover, for soil pollution, both mining operations and the wet beneficiation process have direct impacts. The mining operations process has a direct impact on terrestrial acidification (135,418 metric tons of SO_2 eq per year), and the wet beneficiation process has

[9] The average of 2016-2017: the production of beneficiated phosphate rock was respectively 3.4 MMT and 3.9 MMT, the average production used for the calculation is 3.5 MMT.

a direct impact on terrestrial ecotoxicity (1,785.51 metric tons of 1,4DB eq per year). The wet beneficiation process also has direct impacts on water pollution (92.3 metric tons 1,4DB eq and 53.9 metric tons of P were emitted to water) and water depletion (around 12 Mm^3 per year).

6.7.2 Main life stage contributor

This section aims to identify the main life stage contributors to each environmental impact category and to identify the significant environmental issues. The details on each life cycle stage's environmental impacts are presented in the Annex D, Table D-20.

Figure 6.11 shows the overall share of the three life cycle stages (mining operations, wet beneficiation, and drying) of beneficiated phosphate rock production wet and dry. Mining operations have the largest share in the impact categories PM10, Photochemical oxidant formation (Ozone formation), and Terrestrial acidification for both wet and dry beneficiated phosphate rock. The wet beneficiation process has the highest share in toxicity (for human, terrestrial, and water) and water depletion for both wet and dry beneficiated phosphate rock. When the drying step is included, it has the highest share in climate change (45%), ionizing radiation (46%), and natural land transformation (62%).

For the product system "Wet beneficiation phosphate rock, as P_2O_5-TN", Figure 6.12 shows the direct contribution of the principal two life stages (Mining operations and Wet beneficiation) in comparison to their background processes contribution for each environmental impact categories. The mining operation step contributes the most to Ozone formation, terrestrial acidification, PM10 emission, and climate change. The wet beneficiation step contributes most to terrestrial ecotoxicity, water eutrophication, and water depletion (the value in comparison to the background process is small, but the impact is not negligible).

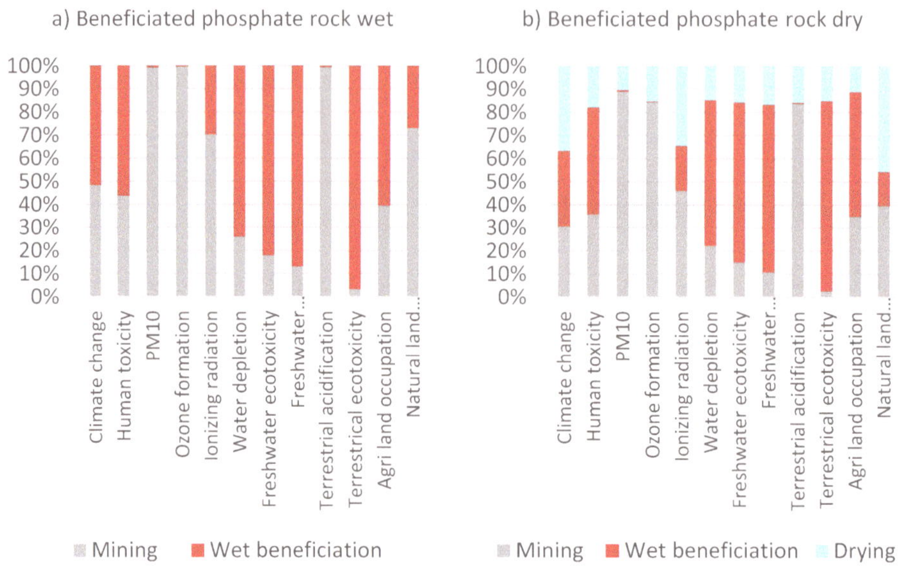

Figure 6.11: Contribution share of each life cycle stage of the production of the beneficiated phosphate rock wet (a) and dry (b)

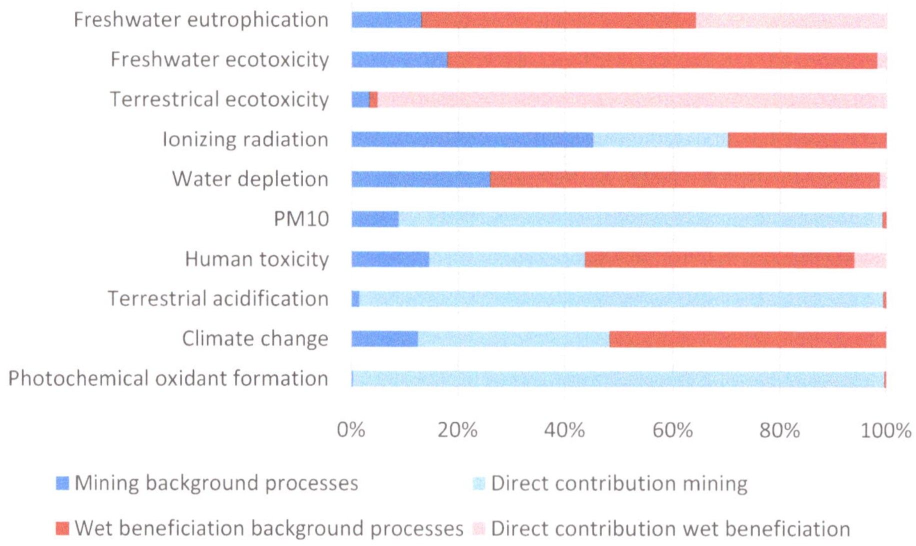

Figure 6.12: Contribution analysis of the mining operations and wet beneficiation in different environmental impact categories

6.7.3 Main key processes contributor

The purpose of this section is to identify the key process contributing to the environmental impact identified as the most significant issues, within a focus on the life stage mining operations. More details about the environmental impact contribution of different operations in the process "Mining operations phosphate rock as P_2O_5-TN" are presented in the Annex D, Table D-21.

The same contribution and gravity analysis were conducted. For mining operations, three clusters of activities were defined; (1) the blasting, (2) the material handling (extraction crushing, sieving, loading, and unloading), and (3) internal transport (Transport of crude phosphate rock and waste backfill). For the material handling, the value is the sum of the contribution of the module "Diesel burned in agriculture machinery" as diesel is consumed during extraction, and the direct contribution of the module "Mining operations phosphate rock, as P_2O_5-TN" as extraction, crushing and sieving release emissions to air (refer to the output Table 6.5). The results of the share of each activity are shown in Figure 6.13. The blasting operation has the highest share for PM10 (75%), Photochemical Ozone Formation (99.9%), and Terrestrial acidification (99.9%). Material handling has the highest climate change (45%) and ionizing radiation (78%). Internal transport has a small share only in ionizing radiation (8.22%).

The gravity analysis results show that blasting generates the highest environmental impact, which is the Photochemical Ozone Formation. It is also responsible for terrestrial acidification. For instance, during blasting, a large volume of Nitrogen oxides NO_x is released, which is increases the potential of ozone formation and terrestrial acidification. Moreover, the use of explosive lead to the production of dust. The quantity of Toxic gas emission and the total suspended particle TSP of the module "Blasting phosphate-TN" are in Table 6.4. The effects due to the use of explosives are located in Tunisia. However, the conditions of explosive production are in Switzerland as the module "Explosive production, Tovex" was used for the blasting process. Material handling is responsible mainly for climate change due to the high consumption of diesel during the extraction and dry mechanical treatment.

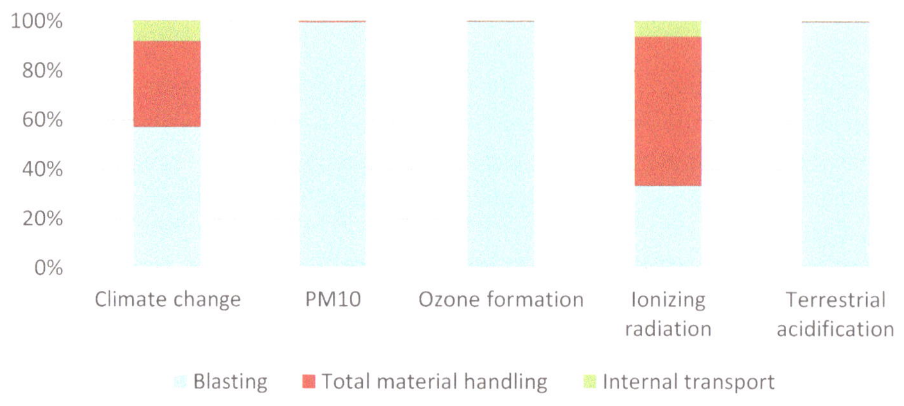

Figure 6.13: Contribution of different operations of the process "Mining operations phosphate rock, as P_2O_5-TN"

6.8 Comparison to the beneficiated phosphate rock production system in Florida, USA

This section aims to assess the variability of the environmental impacts by comparing the results of the current study to a similar product system of beneficiated phosphate rock. Therefore, the Tunisian production system was compared to a production system in Florida, USA. The two investigated product systems have comparable process steps and technologies (see Table 6.12). The product system of beneficiated phosphate rock in Florida was taken from the Ecoinvent database using the same LCIA method (ReCiPe, Midpoint (H)). It is based on the LCA study conducted by Primas (2007) [159].

Table 6.12: Characteristic of beneficiated phosphate rock production in Tunisia and the USA

	USA (Florida)	Tunisia
Final product	Wet beneficiated phosphate rock at 29%	Wet beneficiated phosphate rock at 29%
Mining operations	open pit	open pit
Ratio benef. PR: moved soil	1:11	1:23
Resource in ground	Phosphorus 18% in apatite, 4% in cure ore in-ground	Phosphorus 18% in apatite, 11% in cure ore in-ground
Beneficiation	Washing with water + Flotation	Washing with water

The results of the comparison show the same environmental impact variability profile. This to say, the production of wet beneficiated phosphate rock has a high impact on water resource depletion, the emission of GHGs, and the exposure of humans to radiation and cadmium (see Figure 6.14). Moreover, there is a similarity between the results for the impact category water depletion, water ecotoxicity, and terrestrial ecotoxicity. The product system in the USA (Florida) has a higher radiation impact as the emission of uranium during mining (5.86 E-03 kBq/kg P_2O_5) is higher than in Tunisia

(4.06 E-03 kBq/kg P_2O_5) and higher eutrophication potential. The detailed results of the comparison can be found in Table D-22 in Annex D.

The results of the E-LCA show that the beneficiation of phosphate rock has a direct depletion contribution of 0.012 m^3/kg P_2O_5. The calculated input of new pumped well water per kg beneficiated phosphate rock is 0.0034 m^3 (0.0117 m^3 per kg P_2O_5). Although the company recycles 51-55% of the total water use, recycling can be improved. For instance, in phosphate mining in Florida, the average recycling of water is 95.6%. The average pumped water needed per kg beneficiated phosphate rock is 0.0029 m^3 (0.01 m^3 per kg P_2O_5) [159].

The product system in Tunisia has a higher climate change potential because of the higher consumption of diesel. While in Florida, electric power is used for material handling and conveyor transport of crude phosphate rock, the mining operations in Tunisia are performed by diesel running machines, and transport is done by trucks. Besides, the high ratio of extraction (beneficiated phosphate rock to total moved soil) increases the need for more power to recover the phosphate layer between high thick topsoil and overburden.

Moreover, the product system in Tunisia has a higher impact on human exposure to hazardous elements (Cd, PM10, and photochemical ozone formation). This is explained as blasting is practiced in Tunisian mining sites (blasting is the most significant contributor to Ozone formation and PM10). However, no blasting was reported in the LCA study on Florida phosphate rock. In addition, in Florida, the crude phosphate rock is ground wet without releasing emissions to air. The higher potential of human toxicity in Tunisia is related to the higher content of cadmium (8 ppm in the beneficiated phosphate rock from Florida vs. 35 ppm in the beneficiated phosphate rock from Tunisia).

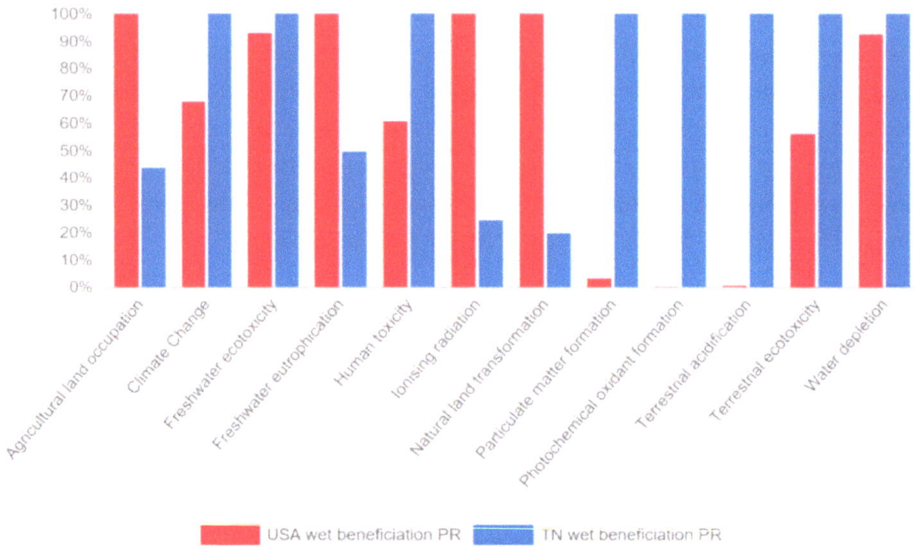

Figure 6.14: Comparison of the environmental impacts of two product systems of wet beneficiated phosphate rock in Tunisia and the USA – Relative indicator results of the respective product systems. For each indicator, the maximum result is set to 100%, and the results of the other variant are displayed in relation to this result

6.9 Conclusion and Recommendation

In conclusion, the production of beneficiated phosphate rock in Gafsa has an impact on the three studied areas of damage mentioned in the scope of the study. The mining operations are the main contributor to the damage to human health and soil acidification due to the use of nitrate explosives. The wet beneficiation process is the main contributor to water depletion and water eutrophication, and for the dry beneficiated phosphate rock, the use of heavy fuel oil is the main cause of GHG emissions. The climate change potential of the dry beneficiated phosphate rock is almost double of the wet beneficiated phosphate rock. As a recommendation, the use of explosive has to be monitored closely, and further investigation is recommended as blasting is the main process contributor to many environmental impacts.

In order to reduce the climate change impact, energy consumption during the mining operations has to be reduced. The main energy source is diesel. Reducing diesel consumption can be achieved by optimizing the transport between sites and the washing plants. As the company has a proper infrastructure for transport by conveyor and train, a follow up comparative LCA study would be valuable to compare different transport scenarios as well as comparing the different types of energy source.

7 Social Life Cycle Assessment of phosphate rock mining in Gafsa, Tunisia

7.1 Introduction

In order to achieve a sustainable global supply chain of minerals, it is important to know where and how the minerals used have been affecting human societies. Besides the environmental impacts generated during the mining of mineral resources, the social impacts of the mining sector are at the heart of the discussion about the sustainability of the supply chains of industrial products. Therefore, the value of minerals on the global market must reflect its extractive economic, environmental, and social costs [100].

In the previous Chapter 6, the E-LCA method was used to investigate the environmental pressure of phosphate rock mining in Tunisia. Similarly, the S-LCA method will be used to measure the social impacts through the life cycle of phosphate mining in Tunisia.

In this chapter, the approach to investigate the social impacts and the social value of phosphate mining through its life cycle is described in detail. Afterward, the results of social LCI are discussed. Besides the S-LCA method, social impacts in Tunisia were investigated by conducting a survey in the region to have insights about the perception of the local population on the impacts of the phosphate rock mining activity. Finally, the results of the S-LCA were compared to the results of the survey among the local population.

7.2 Goal and scope of the study

7.2.1 Goal

In a climate of regional social turmoil [143, 172], the S-LCA aims to uncover the social impacts related directly and indirectly to the mining and production of the beneficiated phosphates rock in the area. The analysis focuses on the social impact categories, which are related to the socio-political as well as the environmental situation of the region (as discussed in chapter 5), such as local employment, access to health facilities, and water scarcity and pollution.

7.2.2 Scope

The scope of the study is the extraction of phosphate rock in nine different sites, the washing of crude phosphate rock in ten different washing plants, and the supporting services such as the machinery maintenance department, the training center, and the headquarter of the CPG in Gafsa.

The scope is the mining area in Gafsa, south of Tunisia. The scope of the studied system is the same as in E-LCA (see chapter 6, section 6.3). The complementarity of E-LCA and L-CA methods discussed in Chapter 2 will enable the discussion of the ecological and social impact interaction.

7.3 Functional unit and system boundary

7.3.1 Definition of the functional unit

As P_2O_5 is considered a fertilizer unit, and to conserve the alienation with E-LCA, the functional unit is not expressed in monetary value instead expressed in physical value in kilogram (kg). The functional unit used for this study is 1 kg of P_2O_5. The content of P_2O_5 in the beneficiated phosphate rock produced by the company (CPG) is 29%

7.3.2 Definition of the system boundary

The inputs into the system are physical inputs expressed in monetary values (such as water, energy, fuel, and chemicals), the operations and services (such as transportation and workers training), and the work performed by the workers.

The output of the system is the final product, which is 3.4 kg of beneficiated phosphate rock (29% P_2O_5) corresponding to 1 kg of P_2O_5 (the functional unit) and its social impacts, which can be defined by human rights, health, and safety and economic prosperity and be expressed in risk per hour.

7.4 Social impact characterization

The assessment framework is illustrated in Figure 7.1, according to the guidelines of the United Nations Environment Programme (UNEP, 2009). The stakeholders' categories selected for this study are: "Local population", "Employees", "Value chain actors", and "Society". These are the categories used in PSILCA and the UNEP guidelines and show high power to influence the production (see Methodology, section 2.4.1). Although the fertilizer company as a client of the CPG also belongs to the high power stakeholder category. However, they were excluded because they belong to the same company group, GCT.

The subcategories are socially significant attributes classified according to a stakeholder impact category. They are assessed by indicators that can be measured by the unit of inventory data (unit of measurement) [202].

Stakeholder category	Impact subcategory	Indicator	Inventory data
Worker	- Child labor - Forced labor - Etc..		
Local population	- Health and safety - Unemployment - Etc..		
Society	- Governance - Etc..		
Value chain actors	- Corruption - Etc..		

Figure 7.1: Impact assessment framework: from category to unit of measurement based on the guidelines for SLCA of UNEP [202]

7.4.1 Stakeholder categories

The results of the Stakeholders power interest Matrix show that the stakeholders belonging to the high-power/high-interest sector are the state as the main shareholder, the providers of transportation, the workers (white and blue-collar), the syndicates, the local population, the local environmental NGOs, and the banks. These stakeholders need to be closely managed by the company. In contrast, the stakeholders in sector low-interest/high power are the electricity provider and the local and national media. These stakeholders need to be kept satisfied. According to the classification list of stakeholders, society has low power and low-interest in the company. However, after the revolution, Tunisian society did require from the government and the national companies to improve their transparency and apply good governance rules. Social movements across the country were organized on social media platforms as well as on the streets, putting pressure to increase the transparency of the management of natural resources. The new Tunisian constitution of 2014 answered this request by adopting the rules of good governance in articles 12 and 13, and the rules for more transparency in article 15 [97, 98, 180].

7.4.2 Social impact subcategories and indicators

The selection of the subcategories and the indicators in social life cycle assessment is a very sensitive task and a very important step. To this day, there is not a commonly accepted standard for social indicators [29].

For this work, the impact subcategories are the key social issues addressing four stakeholder categories: "Local population", "Employees", "Value chain actors," and "Society". The social impact subcategories were selected based on the PSILCA database [29], the specificity of the mining sector [127, 128], and the guideline of the UNEP/SETAC working group [202]. In addition, two impact subcategories were developed for the case of phosphate mining in the Gafsa region. The two new social impact subcategories are the **socio-political stability** concerning the stakeholder society and the **local economic opportunity** concerning the stakeholder: local community (see Table 7.1).

Both subcategories reflect the role of mining to promote regional economic and social stability.

The socio-political stability is the degree of human development and economic freedom of the society as well as the degree of the political stability of its government (risk of destabilization by unconstitutional or violent means) [118, 223]. It is defined by the Human Development Index (HDI) [200], the World Governance Index WGI [223], and the Index of Economic freedom (IEF) [134].

The economic opportunity is the degree of diversification of the economic activity in a region, which reflects the dependency of the region to a specific economic sector, increasing output volatility and making the regions more vulnerable to shocks [151]. It is measured by the concentration of the economy and expressed by the HHI.

The impact indicators are classified into two classes; the enabling environment indicators and the sector-specific indicators (see Table 7.1). The indicators of the enabling environment are descriptive. They reflect the political situation and the social and environmental performance of the country and the region. They describe how the existence of policies and the applicability of the law in Tunisia in general and Gafsa in specific could affect the social performance of the phosphate mining sector. The sector-specific indicators describe how the production process and management conditions of the phosphate mining sector could affect its stakeholders positively or negatively. These indicators are specific to the phosphate mining in Tunisia.

In order to assess the suitable social impacts that reflect the effect of the phosphate mining sector in Tunisia, the indicators were selected from the PSILCA database for the extractive sector in Tunisia with a focus on the sector-specific social impact indicators. The social impact and their indicators are described in Table 7.1.

7.4.3 Normalization

The Normalization follows the PSILCA database normalization method [29]. Hence, the indicators depending on the size of the sector are normalized by the number of employees in the sector or by its total output. For example, the indicator rate of the fatal accident is normalized by 100,000 workers in the sector. For each indicator, the risk assessment structure is adopted from the PSILCA database scheme for risk level assessment [29].

There are six assessment levels of the risk: no data, very low risk, low risk, medium risk, high risk, and very high risk. Similarly, there are five assessment levels of opportunity: no data, no opportunity, low opportunity, medium opportunity, and high opportunity.

Table 7.1: Assessed stakeholders, impact subcategories, applied indicators, and the data sources

Category	Subcategory	Indicator	Metric (Unit)	Data source
Workers	Child labor	Children in employment female	% Female children ages 7-14	[22]
		Children in employment male	% Male children ages 7-14	[22]
		Children in employment total	% Children ages 7-14	[22]
	Working time	Weekly hours of work per employee	Hour	[56]
	Forced labor	Presence of individual working without a contract or with a contract, which does not respect the working law	% per 1,000 employee	[36, 162]

	Discrimination	Gender wage gap	%	[162] [152]
		Women in the sectoral labor force	Ratio	[56][56] [90] [57]
	Fair salary	Company minimum wage per month	Ratio	[56] [106, 107, 155]
		Company average wage per month	Ratio	[56] [106, 145, 155]
	Health and safety measure	Fatal accident rates at the workplace	Per 100,000 worker/year	[50, 51]
		Non-fatal accident rates at workplace	Per 100,000 worker/year	[50, 51]
		Presence of sufficient safety measure	Per 100,000 worker/year	[49]
	Social benefits, legal issues	Social security expenditure of the country	% of GDP	[113]
Local population	Local employment	Average unemployment rate	%	[147]
	Access to material resources	Level of industrial water use related to water withdrawal in the region	% of total water use in the region	[61] [147]
		Level of industrial water use related to water resources in the region	% of total renewable water resources	[61] [147]
		Certified environmental management system	Per 10,000 employees	[115] [108]
	Safety and healthy living condition	Pollution level of the country	Index	[222]
		Drinking water coverage	%	[105]
		Sanitation coverage	%	[105]
		Contribution of the sector to $CO_{2\,eq}$ load	kg	E-LCA Chapter 6
		Contribution of the sector to NMVOC load	kg	E-LCA Chapter 6
		Contribution of the sector to PM10 load	kg	E-LCA Chapter 6
	Emigration and delocalization	Net emigration rate	‰ per year	[147]
	Economic opportunity	Regional Economic activity concentration	HHI	[151]
Society	Contribution of the sector to the economic development	Add value share of the sector of the GDP	% GDP	[109]
	Natural resources Governance	Natural resources Governance index	Index	[144]
	Socio-political stability	The 'loss' in human development due to inequality	(HDI-IHDI)/HDI	[200]

		Political Stability Absence of Violence, Terrorism	Index	[223]
		Economic freedom	Index	[134]
Value chain actors	Public sector corruption	Public sector corruption perception index	Index	[192]
Subcategory add to the PSILCA subcategories Enabling environment indicators Sector-specific indicators				

7.5 Data collection

7.5.1 Data sources

The data concerning the workers and the local population were collected from the company's human resources reports and activity report for the year 2016-2018, the internal audit report for 2017, and the results of the E-LCA about the environmental load and water consumption for 2017.

The data used for the local population are specific to Tunisia and the Gafsa region. The source of data is the National Statistical Institute (INS) and regional reports. The data concerning the society and the value chain actors are published data from national statistical reports and international reports, such as from the World Bank, the IMF, and the NRGI. Table 7.2 shows the data sources for each indicator. For the upstream processes, the PSILCA database was used.

7.6 Inventory analysis

7.6.1 Activity variable

A common "activity variable" was attributed to all the unit processes. The need to define an activity variable is to express the relevance of the unit process contribution to the product's life cycle.

According to the guidelines for the social life cycle assessment of products, "*the activity variable scaled by the output of each relevant process is used to reflect the share of a given activity associated with each unit process*" [202].

The most common activity variable used in S-LCA is worker-hours. It expresses how many worker-hours are associated with the output (the risk level) of each unit process. That is to say, how many hours workers spend in that operation to deliver the functional unit under certain working conditions and at a certain level of risk. Such information is needed to prioritize unit processes and to redefine the boundaries of the product system if it is needed. Furthermore, it helps to decide whether primary data need to be collected or generic data is sufficient for a given unit process [202].

For instance, all unit processes to deliver 1 kg of P_2O_5 (exploration, mining operations, screening, washing, supporting services, and transport) are located in the geographical scope and performed by the same company (CPG). Therefore, to define the worker-hours for all unit process, on-site data was collected from the company for every unit. The worker-hours are related to 1 USD output generated by the process. The worker-hours are calculated according to the formulas in the PSILCA database as followed:

$$(1)\ \boldsymbol{Worker-hours} = \frac{Unit\ labor\ costs\ (USD)(2)}{Mean\ hourly\ labour\ cost\ per\ employee\ \left({}^{USD}/_{h}\right)}$$

$$(2)\ \boldsymbol{Unit\ labour\ costs} = \frac{Compensation\ of\ employees\ per\ year\ (USD)}{Gross\ output\ of\ the\ company\ (USD)}$$

In 2017, according to the company's accounting report, the labor costs (salaries + extra-hours compensation + bonus) represented 35% of the gross output of the company. The average hourly wage (all worker categories included) is around 6 USD.

As the functional unit is expressed in kg (not in USD), the activity variable was normalized to 1 kg P_2O_5 output. Hence, the worker-hours per 1 USD output were multiplied by the price of 1 kg P_2O_5. According to formula (3), the worker-hours to produce 1 kg of P_2O_5 are 0.013 h.

$$(3)\quad \boldsymbol{Worker-hours} = \frac{Unit\ labor\ costs (USD)}{Mean\ houlry\ wage\ (USD/h)} * price\ of\ 1\ unit\ product\ (USD)$$

$$\boldsymbol{Worker-hours}_{at\ the\ CPG\ per\ year\ 2017} = {}^{0.35}/_{5.4} * 0.20 = \mathbf{0.01307072\ (h)}$$

Indicators concerning the local community, society, and value chain actors are also expressed in worker-hours. According to PSILCA database documentation, more suitable activity variables for the local community, society, consumer, and value chain actors are being developed [29].

7.6.2 Flow Analysis

7.6.2.1 Local population

The results of the LCI for the local population (see Table 7.2) show that the mining region has a very high risk of unemployment. According to the National Statistical Institute (INS), the Gafsa governorate's unemployment rate was 26.3% in 2016 [146]. For the mining area (Metlaoui, M'dhila, Moulares, and Redeyef), the unemployment rate was 34%. However, the national unemployment rate for Tunisia was 14% for 2016.

In addition to the high unemployment rate, there is a very high risk of regional concentration of economic activities. Although the phosphate industry contributes 32% to the local employment[1], the unemployment rate is high due to the absence of other economic opportunities to absorb unemployment. After the revolution of 2011, the company proceeded to hire from the local population and surrounding cities based on social criteria. The number of workers increased from 4,898 in 2011 to almost 7,000 workers in 2016. Despite being a main employer in the region, the CPG could not absorb the totality of the

[1] Contribution of the sector in the local employment= number of employed local population in the sector / total employed in the sector at the country level * 100

unemployment problem in the region. Consequently, the region suffers from a very high risk of emigration; the young population is seeking job opportunities in other Tunisian cities.

Furthermore, the region has a very high risk of sanitation coverage, while the drinking water coverage is of very low risk. According to the local authorities, the sanitation coverage is 69.2%, and the drinking water coverage is 99.5% [147].

The flows concerning the impact of the phosphate industry on the local population show that there is a very low risk of industrial water use to compete with the water withdrawal for the local population. However, there is a medium risk concerning the industrial water withdrawal related to renewable water resources in the region.

The high risk related to the presence of the phosphate industry in the region is the contribution of the industry to air and water pollution, as well as the absence of a certified environmental management system.

Table 7.2: Flow analysis results for the impact category: the local population

Subcategory	Indicator	Value	PSILCA risk scheme	PSILCA flow indicator
Local employment	Average unemployment rate in the region	34%	0% - < 3% = very low risk; 3% - < 8% = low risk; 8% - < 13% = medium risk; 13% - < 18% = high risk; >= 18% = very high risk; n.a. = no data	Unemployment rate in the country, very high risk
Access to material	Freshwater, desalinated water, and treated wastewater withdrawn by the Phosphate industry related to total water withdrawal in the region	1.99%	0 - <10% = very low risk; 10 - < 20% = low risk; 20 - <30% = medium risk; 30 - <40% = high risk; >= 40% = very high risk; n.a. = no data	Level of industrial water use (related to total withdrawal), very low risk
	Freshwater, desalinated water, and treated wastewater withdrawn by the Phosphate industry related to total declared renewable water resources in the region	5.10%	0 - <1% = very low risk; 1 - <3% = low risk; 3 - <7% = medium risk; 7 - <13% = high risk; >= 13% = very high risk; n.a. = no data	Level of industrial water use related to actual renewable water resources, medium risk

	Number of Certified environmental management systems (CEMS) (ISO 14001) in the sector per 10,000 employees in the country	0.53	>=100 = very low risk; 10-<100 = low risk; 1-<10 = medium risk; 0.3-<1 = high risk; 0-<0.3 very high risk; n.a. = no data	Certified environmental management systems, high risk
Safety and healthy living conditions	Pollution level of the country: Environmental Performance Index (EPI)	62.35	0-<25= very high risk; =>25-<50= high risk;=>50-<75= medium risk; =>75-85= low risk; =>85= very low risk	Environmental Performance, medium risk
	Drinking water coverage: the percentage of the population using an improved drinking water source	99.5%	<= 80 = very high risk; > 80 - 85% = high risk; >85% - 90% = medium risk; >90% - 95% = low risk; >95% - <100% = very low risk; 100% = no risk; n.a. = no data	Drinking water coverage, very low risk
	Sanitation coverage: the percentage of the population using improved and safely managed sanitation facilities	69.20 %	<= 80 = very high risk; > 80 - 85% = high risk; >85% - 90% = medium risk; >90% - 95% = low risk; >95% - <100% = very low risk; 100% = no risk; n.a. = no data	Sanitation coverage, very high risk
	Contribution of the sector to environmental load, $CO_{2\ eq}$/kg P_2O_5	3.75E -01 kg	0 – 1E-05 = very low risk; >1E-05 – 1E-04 = low risk; >1E-04 – 1E-03 = medium risk; >1E-03 – 1E-02 = high risk; >1E-02 = very high risk; no data	Contribution of the sector to environmental load, $CO_{2\ eq}$, very high risk
	Contribution of the sector to environmental load, NMVOC /kg P_2O_5	2.43E -01 kg	0 – 1E-07 = very low risk; >1E-07 – 1E-06 = low risk; >1E-06 – 1E-05 = medium risk; >1E-05 – 5E-04 = high risk; >5E-04 = very high risk; no data	Contribution of the sector to environmental load, NMVOC, very high risk
	Contribution of the sector to environmental load, PM10/kg P_2O_5	7.18E -02 kg	0 – 1E-07 = very low risk; >1E-07 – 1E-06 = low risk; >1E-06 – 1E-05 = medium risk; >1E-05 – 5E-04 = high risk; >5E-04 = very high risk; no data	Contribution of the sector to environmental load, PM10, very high risk
Emigration and delocalization	The difference between the number of people entering and leaving the region during the year per 1000 person	23.17 ‰	0 ‰= no risk; 0 ‰-<|2.5|‰= very low risk; |2.5|‰-<|5|‰= low risk; |5|‰-<|10|‰= medium risk; |10|‰-<|15|‰= high risk; >=|15|‰= very high risk; no data	Net emigration rate; very high risk
Economic Opportunity	The regional economic activity concentration: HHI	0.30	<0.1 very low concentration 0.1-<0.15 low concentration. 0.15-<0.18 moderate concentration. 0.18-<0.25 high concentration. >= 0.25 very high concentration	Economic activity concentration, very high risk

7.6.2.2 Workers

The results in Table 7.3 show that the highest risk concerns the health and safety of workers. This is because of the high rate of fatal and non-fatal accidents. Between 2016 and 2017, 3 fatal accidents happened in the mining area and the washing plant. The fatal accident costs the company the productivity of 300 working days. The non-fatal accidents for the same period cost 135 for each year.

The high rate of accidents is explained by the absence of safety measures. In fact, according to the audit report 2017, there is a total violation of safety and health standards by the workers. According to research during the data collection and the state of our knowledge, for the environmental pathologies due to the exposure to phosphate mining, there was no data available in the ministry of public health, the regional hospital, or in the National Statistical Institute. As well, there was no study about the long-term effect of the exposure to phosphate mining on the health of the workers or the local population. The company provides occupational health care services for their workers, but they do not investigate such issues.

For the gender discrimination inside the CPG, the results show no gender discrimination for the declared official salary. The women and men are equally paid according to the wage payment grid fixed by the internal regulations of the CPG. The payment grid in the company is a seniority-based pay scale and qualification-based pay scale [162]. In addition, the gender wage gap indicator, according to PSILCA, compares the median wage of men and women (monthly salary). However, it does not consider the working hours, discrimination toward access to promotion, financial advantages (e. g. bonuses or company cars), and training.

Despite wage equality, the presence of women in the sector is of high risk. The phosphate industry is dominated by men. At a national level, the ratio between the share of women employed in a sector out of the total active female population in a country (0.008%), and the share of men and women working in the sector out of the total active population in the country (0.026%), is 0.3. At a company level, the women share 13% of total employees with a ratio of 1:7 for blue-collar and 1:4 for white-collar workers (junior and senior managers and engineers).

The working conditions inside the CPG, in general, are good. There is a low risk concerning working time. The official working time duration at CPG in the mining area (another time regime was applied for employees in the headquarter in Tunis and the storage center in Sfax) is 48 hours a week and the working hours per day has not to exceed 9 hours according to the company's internal regulation [162]. The internal regulation of CPG applies the working regime mandatory by the Tunisian labor code modified by the Law n° 96-62 [161]. The working hours carried out per worker per week for the year 2018 at CPG were calculated to be 47.2 hours, below the official working time duration. This is explained by an average absenteeism rate of 13%.

Besides, the average salary at the CPG is 2.6 times the national average salary, which is 280 USD (800.65 DNT) per month [145]. The minimum salary at the CPG is even higher than the average national salary and the national living salary. The workers with a minimum salary at the CPG can afford a decent living standard for their families. Moreover, the CPG is one of the most coveted employers in the region and the country because of the financial benefits. For those reasons, forced labor in the company is no risk. By contrast to forced labor, the company was accused of having an unfair hiring process [220].

Table 7.3: Flow analysis results for the impact category: Workers

Category	Indicator	Value	PSILCA risk scheme	PSILCA flow indicator
Child labor	Percentage of all children ages 7-14 in labor	0.2%	0% = no risk; 0%-<2.5% = very low risk; 2.5%-<5% = low risk; 5%-<10% = medium risk; 10%-<20% = high risk; >=20% = very high risk; n.a. = no data	Children in employment; total, very low risk
	Percentage of male children age 7-14 in labor	0.3%		Children in employment; male, very low risk
	Percentage of female children age 7-14 in labor	0.0%		Children in employment; female, no risk
Forced labor	Forced labor per 1000 person: a working person without a contract, or the contract is not respecting the working law of the country	0.0%	0 per thousand = no risk; 0-<4 per thousand = very low risk; 4-<8 per thousand = low risk; 8-<12 per thousand = medium risk; 12-<16 per thousand = high risk; >= 16 per thousand = very high risk; n.a. = no data	Frequency of forced labor, no risk
Fair salary	Company minimum wage per month: country living wage/company min wage	0.35	ratio<0.5 = very low risk; ratio 0.5-<0.8 = low risk; ratio 0.8-<1 and MW>300 USD = medium risk; 1-<1.2 and MW ≥300 USD OR ratio 0.8-<1 and MW<300 USD = high risk; ≥1.2 OR ratio ≥1 and MW<300 USD = very high risk	Minimum wage, per month, very low risk

	Company average wage per month: company average wage/country living wage	3.76	≥2.5 = very low risk; 2 -< 2.5 = low risk ; 1.5-< 2 = medium risk; 1-< 1.5 = high risk;<1 = very high risk; n.a.=no data	Sector average wage, per month, very low risk
Working time	The mean of weekly hours of work per employee in the company	47.21h	40-<48 = low risk; 48-<55 = medium risk; 55-<60 = high risk; >60 = very high risk; n.a.= no data	weekly hours of work per employee, low risk
Discrimination	Gender wage gap: Difference between male and female median wages divided by the higher median wage*100	0.0%	0%= no risk; 0%-<5% = very low risk; 5%-<10% = low risk; 10%-<20% = medium risk; 20%-<30% = high risk; >=30% = very high risk; n.a. = no data; not applicable	Gender wage gap, no risk
	Women in the secotral labor force: the share of women employed in the sector out of the total active female population in the country, divided by the share of men and women working in the sector out of total active population in the country	0.30	1= no risk; 0.8-<1 or >1-1.5 = very low risk; 0.6-<0.8 or >1.5 = low risk; 0.4-<0.6 = medium risk; 0.2-<0.4 = high risk; <0.2 = very high risk; not applicable; no data	Women in the sectoral labor force, high risk
healthy and safety	Number of fatal accidents per 100,000 employees and year	125	0-<7.5 = very low risk; 7.5-<15 = low risk; 15-<25 = medium risk; 25-<40 = high risk; >40 = very high risk; no data	Fatal accident rates at workplace, very high risk
	Number of non-fatal accidents per 100.000 employees a year	2167.63	0-<750 = very low risk; 750-<1500 = low risk; 1500-<2250 = medium risk; 2250-<3000 = high risk; >3000 = very high risk; no data	Non-fatal accident rates at workplace, high risk
	QHSE system violation Number of violations of occupational safety and health standards (=OSHA cases) per 100.000 employees in the sector	100	0-< 0.1 = very low risk; 0.1-< 1 = low risk; 1-< 10 = medium risk; 10-< 100 = high risk; >= 100 = very high risk; no data	Presence of sufficient safety measure, very high risk
Social benefits	Social security expenditures as a percentage of Gross Domestic Product (GDP)	7.79%	0-2.5 = very high risk; >2.5-7.5 = high risk; >7.5-15% = medium risk; >15-20% = low risk; >20% = very low risk; n.a. = no data	social security expenditure, medium risk

7.6.2.3 Value chain actors

For the impact category "value chain actors", the public sector corruption was investigated. The main clients and suppliers of the CPG belong to the public sector. Table 7.4 shows that there is a medium risk of corruption. The Corruption Perceptions Index (CPI) of Tunisia in 2008 was 43, according to Transparency International [192].

After the revolution of 2011, civil society, the media, and the union of workers were able to report corruption in the transport of phosphate rock. The subcontracting of transport for private companies instead of railway transport was triggered by blocking the railway lines to the port and the chemical facilities by the social strikes [3]. This situation was suspected to be a case of corruption, and the strikers were paid to freeze the phosphate transportation by train.

The president of the Tunisian Forum for Economic and Social Rights (FTDES) declared in 2015 that the situation of phosphate transport is a corruption case and called the government to investigate it [193]. In addition, the deputy secretary-general of the General Union of Tunisian Workers (UGTT) pointed out the fact that a deputy representing Gafsa in the Assembly of the representative of people owns 100 trucks used for the transport of phosphate [95].

Table 7.4: Flow analysis results for the impact category: Value chain actors

Subcategory	Indicator	Value	PSILCA risk scheme	PSILCA flow indicator
Public sector corruption	Public sector corruption: Corruption Perceptions Index (CPI)	43	0-19 very high risk, 20-39 high risk, 40-59 medium risk, 60-79 low risk, 80-100 very low risk	Public sector corruption, medium risk

7.6.2.4 Society

According to Table 7.5, the contribution of the mining sector to the Tunisian GDP is very low. In 2017, the added value of mining was 0.5 % of the Tunisian GDP. The added value of the phosphate sector (both phosphate mining and fertilizers production) is 2% of the GDP [109]. In comparison to the situation before the revolution, when the production capacity was 8 million metric tons of beneficiated phosphate rock, and the social situation was stable, the contribution of the phosphate industry was 4% to the national GDP [15].

The overall social and political situation in Tunisia for 2017 is considered high risk according to the Human Development Index (HDI) and the Political Stability and Absence of Violence (WGI_{PSAV}). The internal factors are related mainly to the socio-economic crisis in the mining area and the surrounding cities. According to the report of the Tunisian Forum for Economic and Social Rights (FTDES), in 2018, 9,356 social movements were held across the country. The majority of the protest movements in the different regions were socio-economic: to demand employment, improvement of social conditions, or improvement of public services. Moreover, the mining area in the southwest of the country is ranked the second place of social protests with 1,083 protests in 2018 [80]. The external factors are the risk of terrorism infiltration from the south border of the country from the neighboring country Libya, taken hostage for years by terrorist groups and civil conflicts.

Another indicator that reflects the stability of society is economic freedom. For Tunisia, the index of economic freedom IEF is 55.4, which is considered a high risk for individuals to do business and create wealth [134]. It reflects the entrepreneurial environment in Tunisia, which is not as well developed as the need of the young generation. Hence, the burden on big companies and the administration sector is high to afford jobs. This situation can be described as a vicious circle where the high unemployment rate triggers the anger of the society, that turn into protests, strikes and blocking the production and lead to economic failure and deepen the economic crises.

Table 7.5: Flow analysis results for the impact category: Society

Sub-category	Indicator	Value	PSILCA risk scheme	PSILCA flow indicator, according to the report
Contribution of the sector to the economic development	Shares of breakdown of GDP/ Added value at current prices; % GDP	2%	0-<1 = no opportunity; 1-10 = low opportunity; >10-25 = medium opportunity; >25 = high opportunity; no data	Contribution of the sector to economic development, low opportunity
Natural resources Governance	Resource Governance Index (RGI)	48	<30 very high risk,30-44 high risk, 45-59 medium risk, 60-74 low risk,>74 very low risk	Natural resources Governance, medium risk

Socio-Political stability	The 'loss' in human development due to inequality is the difference between the human development index (HDI) and the Inequality-adjusted Human Development Index (IHDI) divided by the HDI expressed in %	22%	0-5% very low risk;6-11% low risk;12-17% medium risk; 18-24% high risk; >25% very high risk	Human development, high risk
	Political Stability Absence of Violence Terrorism percentile rank (0-100) from by the World Governance Index (WGI)	10-25th	0-10th very high risk, 10-25th high risk, 25-50th medium risk, 50-75th low risk, >75th very low risk	Political stability, absence of violence, high risk
	Economic freedom is the fundamental right of every human to control his or her own labor and property; the Index of Economic freedom (IEF)	55.4	<50 unfree (very high risk), 50-<60 mostly unfree (high risk), 60-<70 moderate free (medium risk), 70-<80 mostly free (low risk), 80-100 free (very low risk)	Economic freedom, high risk

7.7 Impact assessment

7.7.1 Social impacts

The social impact categories are listed top-down in Table 7.6. The most impacted stakeholder is the "Local population" with an impact level between 0 and 4.2 medium risk hours. The highest social impacts on the local population come from the categories 'Contribution to the environmental load', 'Sanitation coverage', and 'Economic opportunity', 'Unemployment' and 'Emigration'. Other relevant social impact categories are 'Certified environmental management system' and 'Unemployment'. The stakeholder group "Workers" is also highly impacted; the impact level lies between 0 and 1.3 medium risk hours. The highest impact categories are 'Safety measures', 'Fatal and non-fatal accidents', and 'Women in the sectoral labor force'.

The impact on stakeholders "Society" and the "Value chain actors" are less severe. The impact level lies between 0 and 0.25 medium risk hours. The highest impact category concerning the "Society" is illiteracy, especially for females. The highest impact categories concerning the "Value chain actors" are "Public sector corruption" and "Social responsibility along the supply chain".

Table 7.6: Social impact categories of phosphate mining in Tunisia

Social Impact	Stakeholder category	Value med risk hour
Contribution to environmental load	Local population	4.19
Sanitation coverage	Local population	1.52
Safety measure	Workers	1.32
Fatal accident	Workers	1.32
Unemployment	Local Population	1.30
Economic opportunity	Local population	1.30
Emigration	Local population	1.30
Public sector corruption	Value chain actors	0.24
Social responsibility along the value chain	Value chain actors	0.21
Illiteracy	Society	0.20
Non-fatal accident	Workers	0.19
Women in the sector	Workers	0.13

7.7.2 Geographical distribution of social impacts

After defining the most relevant social impacts along the supply chain of the reference process, the geographical location of the processes responsible for generating the social hotspot will be examined. The hotspot could be generated by the reference process itself ("Phosphate rock beneficiation as P_2O_5 -TN") and/or by one of its direct processes (see Figure 7.2) and/or by an upstream process.

The supply chain model of the reference process and the location of the direct processes are presented in Figure 7.2. The model shows that the direct processes are mostly concentrated in Tunisia. Besides the mining operations and the beneficiation process, the transport, electricity production, construction, maintenance, and explosive and fuel production are also located in Tunisia. The chemicals (Flocculants) production is located in China. The machinery production processes are located in Germany and the United States. Therefore, the geographical location of the defined social hotspots along the supply chain is concentrated in Tunisia.

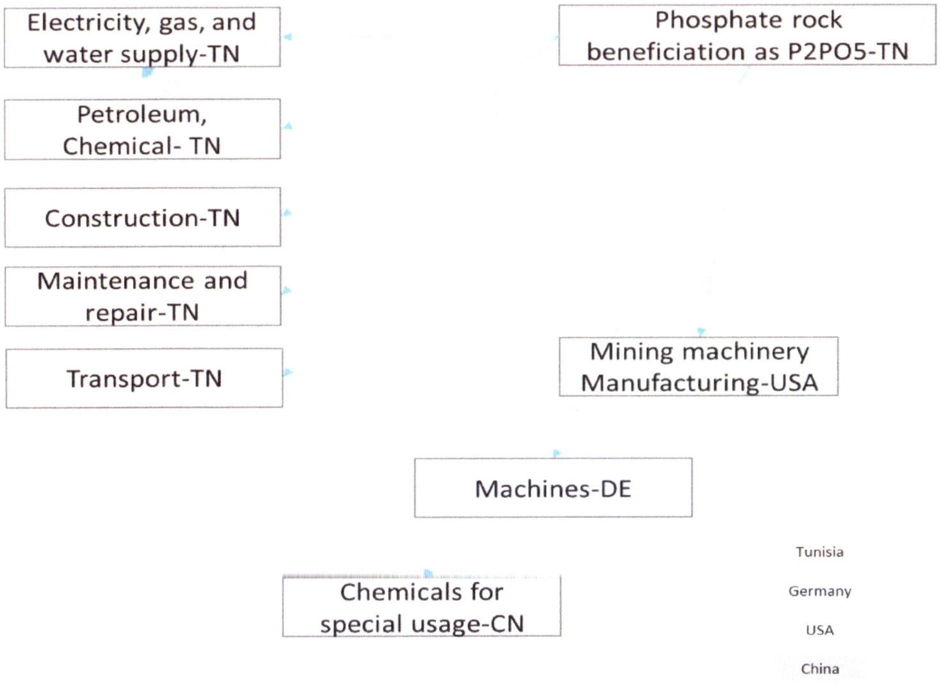

Figure 7.2: Social LCA model of phosphate rock beneficiation as P_2O_5 in Tunisia

For the category "Local population", all the defined social hotspots are localized in Tunisia. The reference process "Phosphate beneficiation as P205- TN" is the main contributor to the social impact category "Contribution to environmental load", "Sanitation coverage," and "Unemployment".

The direct processes like the production of the explosive "Petroleum, Chemical products-TN", the "Transport" and the "Maintenance and repair" are also contributing to social impacts.

For the stakeholder category "Workers", the hotspots are concentrated in Tunisia. In the same way as for the "Local population", the reference process is the primary contributor to the lack of "Safety measure" and the risk of "Fatal and non-fatal accidents". For the impact category "Women in the sectoral labor force", the impact is concentrated mainly in Tunisia. The reference flow is the main contributor in addition to other direct processes, such as construction and transportation. Women are still underrepresented in the sector of heavy industries and mining in Tunisia.

For the stakeholder group "Value chain actor", the public sector corruption lies in Tunisia, and the main process contributors are the "Transport" and "Maintenance and repair". For the stakeholder "Society" the illiteracy problem is located mainly in Tunisia.

7.8 Comparison of the phosphate beneficiation process in Tunisia to PSILCA's generic mining and quarrying product system

In order to compare different product systems, the functional unit must be the same for all the elementary flows. The elementary flow of the product system "Phosphate rock beneficiation as P_2O_5-TN" was recalculated and expressed in 1 USD instead. The results are calculated automatically in openLCA. Figure 7.3 shows the results of the comparison of two product systems: "Mining and quarrying" in Tunisia from the PSILCA database and the product system "Phosphate beneficiation as P_2O_5-TN" in Tunisia from this work.

The chart shows the relative indicator results of the respective product system. For each indicator, the maximum result is set to 100%, and the value of the other product system is expressed in relation to the maximum result.

The results of this study, in comparison to the PSILCA database process, confirm the high risk of non-fatal accidents. However, this study shows that the impact categories "Fatal accidents" and "Safety measure" are higher than in the PSILCA database, and the rest of the impact categories are lower than in the PSILCA database.

The origin of the differences in the results is the size of the sector and, most importantly, the origin of the data. In PSILCA, the data is aggregated for all the mining activities of non-metal minerals in a country, but this study is specific to phosphate mining (Phosphate rock mining is the largest extractive sector in Tunisia [148]). The data origin in the PSILCA database is generic data. For this study, primary data collected from the mining company (CPG) was used as well as data from national and regional institutions.

Concerning the impact on the local population, both product systems have a high impact on the surrounding environment of local communities in Tunisia.

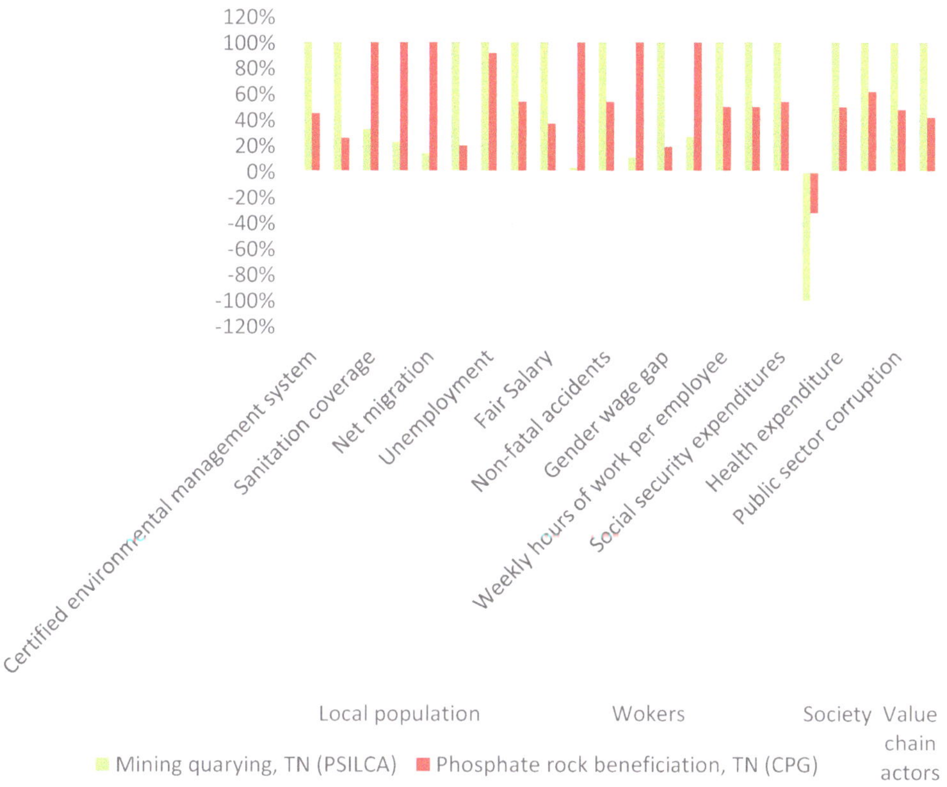

Figure 7.3: Comparison of social impact categories of the process "Mining and quarrying" in Tunisia in the PSILCA database to the process "Phosphate beneficiation as P_2O_5" in Tunisia

7.9 Site-specific social impact assessment survey

7.9.1 Social hotspots definition

The socio-ecological impact results of the S-LCA and E-LCA are used for the site-specific social impact assessment. These social and environmental impacts will be studied further according to the perception of the local population. The main social impact is the health and safety of the local population. It is affected mainly by water and air pollution. According to the S-LCA results, a second social impact is the access to water resources for daily life usage and irrigation in agriculture. A third social impact is the economic opportunity, as the S-LCA results show that the concentration of the economic activity in Gafsa leads to reduced economic opportunities in the region.

The fourth social hotspot, according to the E-LCA and S-LCA results, is the contribution to environmental load. It occurs due to emissions of CO_2 eq, radioactive elements, and particulate matter in the air. Also, the impact on the quality of life and the surrounding environment were investigated. Finally, the involvement of the local population and the degree of being informed by the company were explored.

7.9.2 Survey results

The number of respondents from the local population is 112, with 97 online responses and 15 street-survey responses. The structure of the respondents of the survey is shown in Figure 7.4 and Figure 7.5. The majority of the respondents belong to the young groupage range (20-39 years old). Around 8% more young men answered the questionnaire than young women. Overall, the composition of the sample is homogeneous. The respondents' sample can be considered as representative to the local population, which is composed of 49.9% male and 50.1% female as well as 23% of youth between 20 and 29 years old, 22% of young adults between 30-39 years old, 16% of adults between 40-59 years old and 13% of seniors (>60 years old) (see Annex B, Figure B-3).

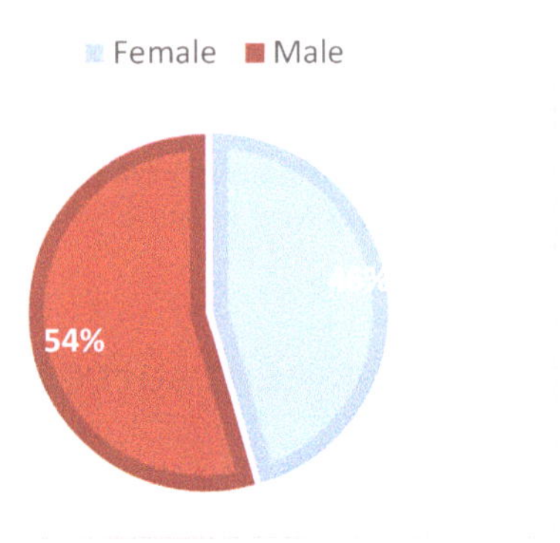

Figure 7.4: **Gender structure of the local population respondents to the survey**

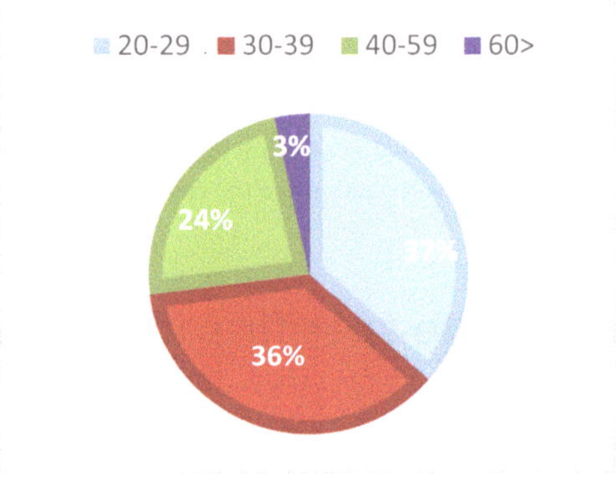

Figure 7.5: **Age structure of the local population respondents to the survey**

7.9.2.1 Health

As a reason behind leaving the region, the low quality of health care was reported by many respondents while conducting the street survey. Therefore, the “Quality of health care” was added to the question: “Can you rate the reason behind leaving the region? Unemployment, Health/ access to clean water/Pollution/Security, Other reason” in the online survey.

The answers to this question show that “Poor quality of health care” is one of the two main reasons for emigration after unemployment (see Figure 7.6). 41% of the respondents think that the quality of health care has a very high impact on emigration, and 35% think that it has a high impact on emigration.

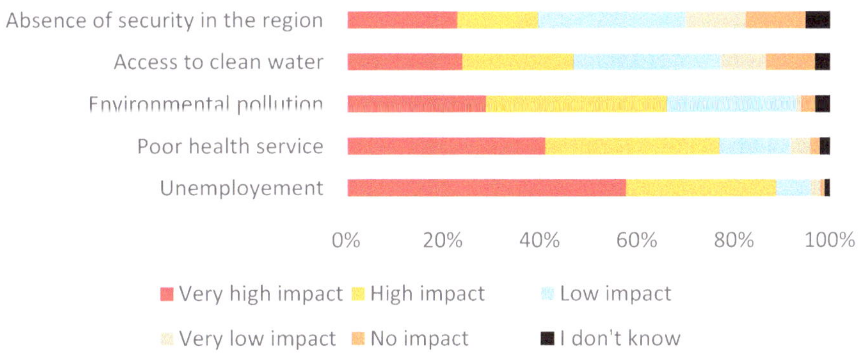

Figure 7.6: Reason behind the emigration according to local population respondents

According to the respondents, there is a very high risk of diseases due to the phosphate industry (see Figure 7.7). 68% of the respondents believe that there is a very high risk of diseases related to air pollution with the dust during the mining operations, and 59% believe that there is a very high risk of diseases related to water pollution and contamination of water by fluorine and other contaminants due to phosphate beneficiation. For the local population respondents, death due to cancer is higher than the national average. 70% of the respondents believe that there is a very high risk of having tumoral diseases due to phosphate mining activities.

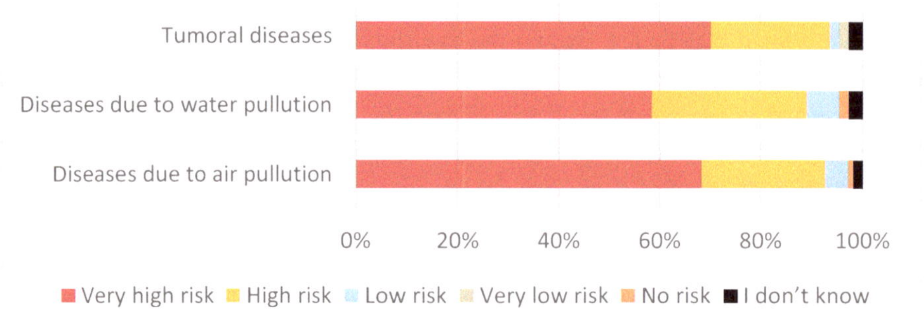

Figure 7.7: Perceived risk of diseases due to the phosphate industry in the region according to local population respondents

7.9.2.2 Access to water sources and quality of water

- **Quality of drinking water in households**

According to the regional report on infrastructure statistics for 2016, 90.7% of the local population is connected to the drinking water network, and 99.5% of the local population has access to drinking water [147]. However, the connection to a source of drinking water is not a sufficient indicator, the quality of water and the interruption frequency in the household are important to assess the quality of access to drinking water. The results in Figure 7.8 show that 53% of the local population respondents have reported that the drinking water in their households has very bad quality, and 26% have reported that it has bad quality.

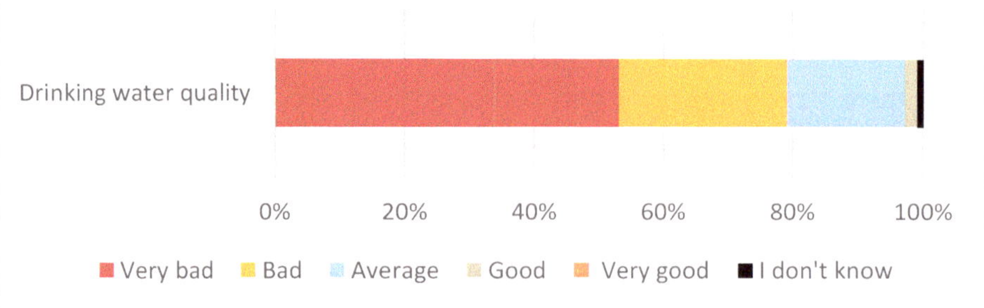

Figure 7.8: Perceived quality of drinking water according to local population respondents

- **Drinking water interruption in households**

The results in Figure 7.9 show that 23% of respondents have reported a very high frequency of water interruption in their households (at least once a day), and 26% have reported a frequent high interruption of water (1 to 3 times per week). However, 35% said that the interruption of water in their households is low frequent (1-3 times per month).

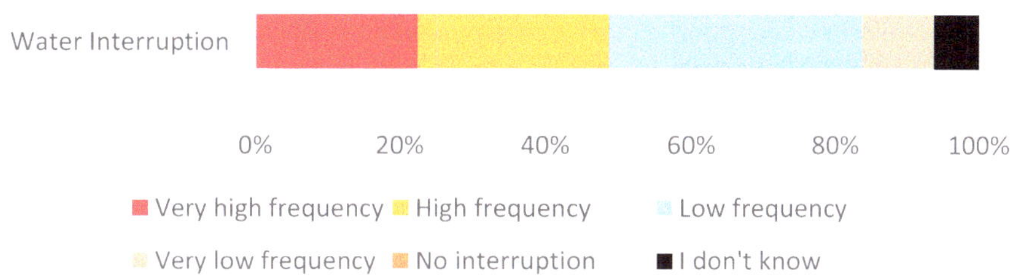

Figure 7.9: Frequency of water interruption in the households according to local population respondents

- **Water source depletion and pollution**

For water depletion, the results in Figure 7.10 show that 59% of local population respondents believe that the phosphate industry has a very high risk of water resource depletion. The results of the S-LCA show that freshwater, desalinated water, and treated wastewater withdrew by the phosphate industry related to total water withdrawal in the region (for agriculture industrial and municipal water) is very low risk.

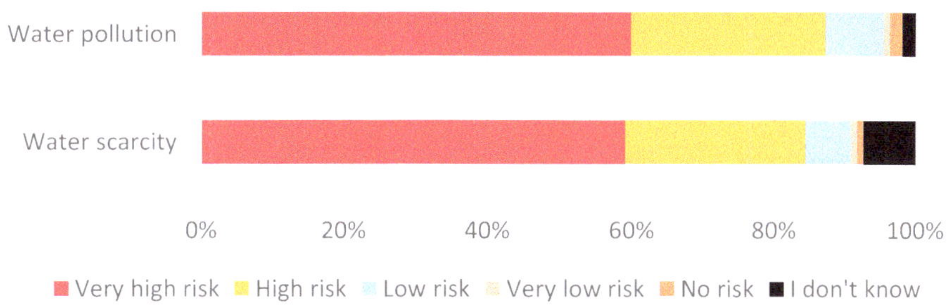

Figure 7.10: Perceived impact of the phosphate industry on the water depletion and water pollution in the region according to local population respondents

Nonetheless, the withdrawal in comparison to total declared renewable water resources in the region is medium risk. The same results apply for water pollution by the phosphate industry (see Figure 7.10). 60% of local population respondents believe that the phosphate industry is polluting the freshwater and impacting the quality of drinking water and the water for agriculture.

7.9.2.3 Economic development

The local economic development impact category refers to the impact of the phosphate industry on the other economic sectors in the mining area and the attractiveness of the region for investments. The results in Figure 7.11 show that 33% of the respondents think that the phosphate industry negatively impacts the overall regional economic development. As well, agriculture is among the most impacted economic sectors according to their answers, as 43% of the respondents believe that the phosphate industry has a very high negative impact on agriculture.

Concerning the attractiveness of the area to investments, 19% of the respondents said that the phosphate industry has a very negative impact, and 23% said it has a negative impact on the image of Gafsa to investments due to pollution, bad transport infrastructure, and social strikes. However, 30% of the respondents believe that the phosphate industry is impacting the attractiveness of the region to investments neither negatively nor positively.

Moreover, the economic opportunity in the mining area is very concentrated according to the perception of the local population respondents. According to the regional statistics on the employment per economic sector for 2016, 1,337 individuals work in the agriculture and farming sector comparing to 7,749 individuals working in the mining sector [147]. However, other economies are well developed in the mining area, such as the construction sector (4,746 individuals) and the service sector (24,195 individuals).

However, the regional statistics aggregate all the self-employment jobs under the service sector, like an ambulant vendor, a vendor in a grocery shop, or an artisan. There is no specification about the working number in the hospitality industry, the finance sector, and the health care sector. Also, there is no specification if the employed individuals in agriculture, construction, and service have the benefit of social security and fair salary. Therefore, the local population believes that the dominance of the phosphate industry in the region is responsible for the weak development of other economic sectors. Hence, there is no alternative to the phosphate industry, which could offer proper employment conditions, especially for youth with a university degree.

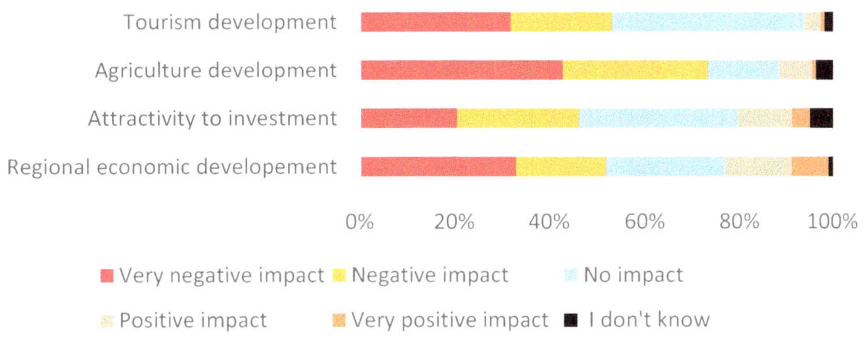

Figure 7.11: Perceived impact of the phosphate industry on the local economic development according to local population respondents

Due to the weak economic development in the region, the unemployment rate is perceived as the first reason for youth emigration (see Figure 7.7). According to Figure 7.12, 62% of local population respondents believe that the high unemployment rate in the region leads to seeking opportunities outside of the mining area. According to regional statistics on employment for 2016, the average unemployment rate in the mining area is 34% [147]. According to the same source, the average net emigration rate per 1,000 persons is 23.173‰. Also, the regional statistics reported that emigration due to professional reasons (unemployment or change in work position) is 23.74%, which represents the third most crucial reason after marriage and family.

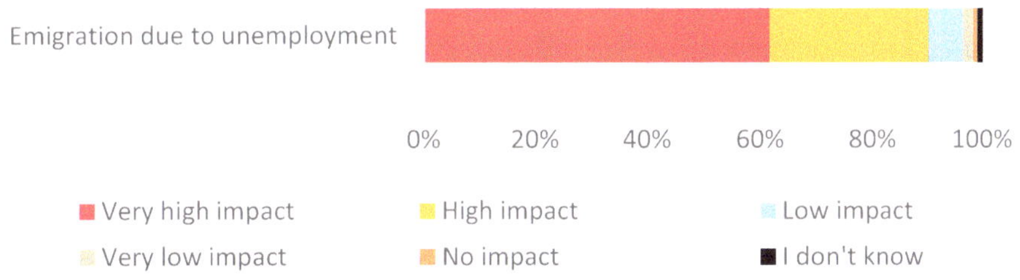

Figure 7.12: Perceived emigration risk due to unemployment in the mining area according to local population respondents

7.9.2.4 Environmental load and quality of life

For the impact of the phosphate industry on the environmental quality and the livability of the region, the results in Figure 7.13 show that 61% of local population respondents believe that there is a very high risk that the phosphate industry reduces the livability of the region. 36% of respondents believe that the phosphate industry has a very high risk of reducing the habitation standards of the area, and 36% believe that it has a high risk due to pollution and the use of explosives, which causes damages to their houses. Moreover, 54% of respondents believe that the phosphate industry has a very high risk of causing the pollution of their arable land.

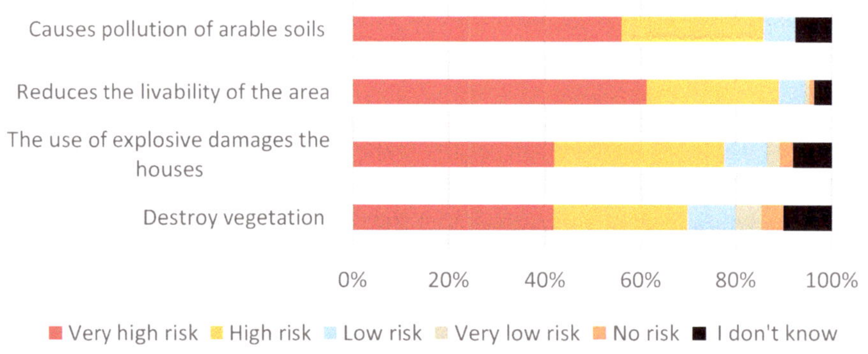

Figure 7.13: Perceived impact of the phosphate industry on the environmental quality and the livability of the area according to local population respondents

7.9.2.5 Local population involvement

Although the local population is recognized as a strong stakeholder by the participants to the stakeholders' Power/Interest Matrix (see Annex B, Figure B-2), 82% of the local population respondents said that they do not participate in any decisions concerning the phosphate activities in the region (see Figure 7.14). Besides being not involved, only 8% of the respondents declare that they are well informed about new mining projects in the region, and 48% claim that they are not informed at all. The same tendency can be seen for the category 'being informed about the social projects financed by the CPG in the region' (see Figure 7.15). Furthermore, 50% of the respondents declared that they are not informed about any health effects; only 9% of the respondents declared that they are well informed about the health effects.

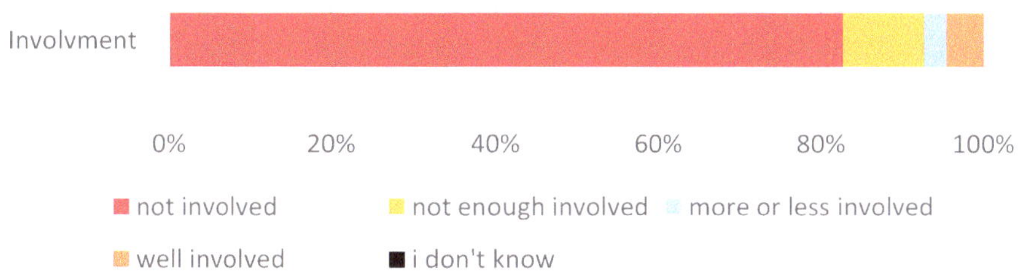

Figure 7.14: The involvement of the local population in decisions concerning phosphate activities in the region according to the local population respondents

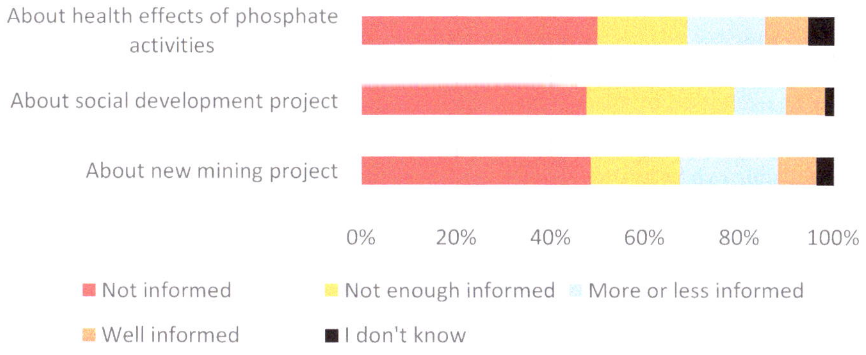

Figure 7.15: The degree of the local population being informed by the company and the local authorities concerning phosphate activities in the region according to the local population respondents

The results of these two last questions show that there is poor communication between the company and the local population. Furthermore, the fact that the local population is not informed and not involved shows that the opinion of the local population regarding other issues (such as water pollution and water depletion, unemployment, health problems) are not channeled to the company which is capable and responsible for answering the local population's concerns.

7.10 Interpretation and discussion

The results of both the S-LCA and the survey among the local population are shown in Table 7.7. It is necessary to compare what the local population believes with the results of the S-LCA. It describes the gap between the reality reported by the phosphate industry and the regional authorities and the reality reported by the local population. The gap becomes apparent for two impact categories; **industrial water use** and the **health of the local population**.

For the first, the results of the S-LCA show that the level of water depletion expressed by the industrial water use related to actual renewable water resources is at **medium risk**. However, the results of the survey are more pessimistic, assuming a **very high risk**. The local population lives in conditions of a high frequent interruption of drinking water and a high risk of water resource depletion for their agricultural needs, which leads to blaming the phosphate industry.

This gap can be explained by the fact that the indicator in PSILCA is based on the level of industrial water related to **renewable water resources**, and the risk assessment considers the industrial water use between 3 and 7% as a medium risk. Though the area has an arid climate, where the rain is scarce (203.7 mm per year [147]), the evapotranspiration is high, and the hydrological resources are depleting very fast, in addition to the contamination of freshwater by saline water due to overexploitation can be seen [94, 173]. Therefore, in the studied area, water resources are considered very low renewable [94].

The risk assessment scale in PSILCA is based on the definition of water stress, which occurs when the withdrawal exceeds 40% of the water resources, according to Bates and Kundzewicz (2008) [11]. It assumes as well that the industrial water use represents around a third of total withdrawal so that the proportion of industrial water withdrawal of total water resources should not exceed 13% [29]. However, the same source mentions that in a global-scale assessment, water stress can also be assumed if the water availability is below 1,000 m^3 per capita per year required for domestic, industrial, and agricultural water usage [11].

In the case of Gafsa, water availability is 684.19 m^3 per capita per year [147]. As such, the region is already water-stressed. The withdrawal of 5% by the phosphate industry alone (not the entire industrial sector in the region) can not be without consequences and puts more pressure on the water resources in the region. According to the E-LCA results, the production of 1 kg of P_2O_5 is responsible for the depletion of 0.012 m^3 of water in the region (direct impact; see section 6.7.1). The average volume consumed by the company between 2016 and 2017 was around 11.8 Mm^3 for an annual average production of 3.5 MMT beneficiated phosphate rock (3.4 m^3/metric ton beneficiated phosphate rock).

The water problem in the region is a common issue for the phosphate industry, other industries, the agricultural sector, and households. As the pressure on the groundwater is increasing, the company is studying the possibility of using desalinated seawater for the beneficiation process. The use of desalinated seawater for industrial purposes can reduce the pressure on the available water resources. However, the mining region needs an inclusive solution, which must be thought out in a way that not only meets the needs of the phosphate industry but also meets the needs of households, the agricultural sector, and farming in the region.

For the health problems, the gap is caused by the absence of data for the S-LCA to either confirm or deny the perception of the local population. The data concerning death due to cancer in Gafsa was investigated to compare it with the perception of the local population. Unfortunately, the register of cancers for the south part of the country reports only the statistic of Sfax governorate for the year 2000-2002 [179].

However, a holistic study was conducted in 2015 by the National Institute of Public Health (INSP) investigating the causes of death in Tunisia [92]. The study has reported that cancer is the second most frequent cause of mortality in the south-west (Gafsa governorate, Tozeur governorate, and Kebili governorate). It causes 11.8% of the deaths, while the national average mortality due to cancer was 16.8% in 2013 [92]. Cancer is also the second most frequent cause of death in Tunisia [92] [224]. The absence of specific data concerning the mining area made the discussion of the local population's perception difficult.

The water pollution is confirmed by both the S-LCA and the local population's perception. The washing of phosphate contributes to the pollution of water resources according to the results of E-LCA. Fluorine is an example of a hazardous chemical element for human health, which was found in water pollution. The discharged water from the beneficiation plant at M'dhila contains around 47 mg/l of Fluorine in the wastewater, which is over the limit fixed by the Tunisian decree 2018 for wastewater discharged in natural water bodies of 3mg/l of fluorine [137]. The occurrence of the high-fluoride groundwater is affecting the safety of the drinking water of the local inhabitants.

Concerning the low economic opportunity coupled with the high unemployment rate in the region, both the S-LCA and the local population's perception show that the economic activity is very high in the region. According to the OECD, the southwest region has a few enterprises operating in the same sectors. In addition, economic activity is highly concentrated in primary products [151]. This high concentration of economic activity increases the vulnerability of these regions to shocks.

For the environmental load and the livability of the area, the S-LCA and the local population perception both agree that the phosphate industry negatively affects the quality of water, soil, and air. Such pollution reduces the livability of the area and negatively affects the image of the area.

Table 7.7: Comparison of the results of S-LCA and the site-specific social impacts (local population perception)

Social hotspot	Perception of the local population	S-LCA results
Health of the local population	Very high risk	No data
Economic opportunity and Unemployment	Very high risk	Very high risk
Freshwater depletion	Very high risk	Medium risk
Water pollution	Very high risk	High risk
Environmental load: soil pollution, noise pollution, the livability of the area, destroy native vegetation	Reduce livability of the area, high risk of noise pollution, very high risk of soil pollution	Very high risk

7.11 Conclusion

The S-LCA model shows that the social hotspots of the production of beneficiated phosphate rock in Tunisia is very concentrated in the mining region, and the reference process is the main contributor to the social impacts. The production processes are impacted by the Tunisian socio-political situation and the regional social stability of Gafsa. Reciprocally the phosphate industry is affecting the Tunisian economy and the regional socio-economic and environmental situation in Gafsa. Furthermore, the S-LCA shows that phosphate mining in Tunisia is not bearing the issues of child labor nor forced labor. The activities of phosphate mining in Tunisia are free from the abuse of human rights. However, the socio-environmental performances are still weak in some areas, like the safety of workers and the health of the local population.

This study shows complementarities between the perception of the local population and the results of S-LCA, but it also shows a difference in the results concerning the pressure on freshwater. Conducting an S-LCA using only the PSILCA database and the related social weighting method cannot capture the complete image of social conditions where workers and local populations live. It is essential to combine it with a site-specific social assessment using qualitative analysis, which can reveal the opinions, feelings, and perceptions of the target group.

The results of S-LCA show a medium risk in the freshwater consumption category. Nevertheless, the local population accuses the phosphate industry of reducing their access to good quality of drinking water. The water problem is general as the region suffers from a shortage of water resources due to climate change.

Besides the water problem, high local unemployment is very problematic for the company. The company and the state address the problem by hiring more, which is never considered enough for the local population. The reason behind unemployment is the high concentration of economic activity in the region and the non-diversification of economic opportunities. The region is rich not only with phosphate but with a long history of 5 thousand years of civilization. The city is also on a touristic road.

However, the tourism sector is absent from the city. Besides, the water problem is jeopardizing the agricultural sector. The region is well known for its oasis and production of dates and grenadines. However, the agricultural activity is in danger due to the increasing emigration of the youth.

Moreover, the results of S-LCA and the survey show – in contradiction to the Stakeholder's matrix developed by the participants from the company – that the regional agencies of tourism and agricultural development are important stakeholders with high power as they can leverage the capacity of the region to absorb unemployment and offer alternative economic opportunities for the local population.

Also, the sanitation provider and the water provider are important stakeholders who could work closely with the CPG to afford reliable and specific studies on water resources in the region, water pollution, water distribution, and the sanitation network. The company has to actively engage all their stakeholders and tackle the different environmental and social issues as a common challenge.

They also need to adopt a participatory strategy involving the local population in decision-making, which would facilitate its work [168]. The results of the survey show that the local population is neglected as a strong stakeholder of the company, and they are not appropriately informed to avoid any misinformation from other parties who can profit from the non-transparent environment to fuel social anger. The social instability can be mitigated by improving access to information, establishing transparency about environmental performance and communication about the company, as well as the socio-economic role of the company as a driver of economic development in the region.

8 Discussion

8.1 Introduction

Phosphorus is a crucial element in agriculture to feed the fast-growing global population. The need to ensure a sustainable supply of phosphorus is above the economic importance of the material. A sustainable phosphorus supply has a social and human dimension as phosphorus serves a basic need, nutrition.

In the first chapter, the research question revolves around the risks and barriers to achieve a sustainable phosphorus supply on the global market. After addressing the concept of phosphorus criticality and the supply bottlenecks of phosphorus on the global market, a link was established between the social and the environmental conditions in the mining region, suggesting that the environmental impacts could be related to the social instability in the region. Thus, this chapter closes the central hypothesis of this thesis with a discussion on the role of the socio-environmental impacts of phosphate mining to mitigate phosphorus criticality.

8.2 Phosphorus criticality and the implication of the social and environmental situation of mining regions

assessment of the time-evolution of phosphate rock reserves, it was found that the global reserves are dynamic. They increase due to the discovery of new deposits as well as due to enabling factors such as technology [175]. Third, the consumption growth rate is twice the growth rate of the reserves. Those issues reveal that despite the abundance of phosphate rock resources, the demand pressure on the available stock is high, which leads to the decline of the ore grade quality. Moreover, other problems emerge. As the ore grade declines, the concentration of cadmium in the final product increases, as well as the consumption of water and energy during the mining and beneficiation process (the lower the P_2O_5 content is in the rock, the more the recovery and enrichment of phosphate rock from the ground consumes water and energy).

Chapter 4 addresses the concept of phosphorus criticality, focusing on supply risk factors. The supply risks were analyzed according to the phosphorus functionality[1] along its global value chain. It was found that the geo-concentration concerns not only the phosphate rock reserves but also the production of phosphoric acid and white phosphorus. Nevertheless, the geo-concentration is higher at the upstream of the supply chain, which is the production of beneficiated phosphate rock. Consequently, the risk of a supply disruption at the production of beneficiated phosphate rock would have a bullwhip effect on the rest of the supply chain. The social stability and environmental performance of the producing country were identified as the main supply risk factor, which could increase the criticality of phosphorus globally.

The social instability in phosphate producing countries seems to be more local than national. Usually, the social conflicts are more located in the mining region (except in the case of Syria, where the war did not start initially in the mining area, and as a consequence, the production and exportation of phosphate were heavily impacted). For instance, the majority of global reserves are found in the disputed region of Western Sahara, controlled by Morocco [142]. The political conflict in Morocco, like the natural resource issues tangled into it, will have major issues for future food availability, especially in the developing countries, as the mining region could have a global monopoly over phosphate [117]. Also, the socio-political conflict in Tunisia started in the mining region (see chapter 5). Since the events in the mining region in 2008, the local population claims their rights for better socio-economic equity [8, 9, 125].

The Resources Governance Institute (RGI) indicates that a country's resources governance capability is an indicator that is more specific to the situation of the mining sector in the mining countries. The indicator encompasses not only the enabling environment, such as the socio-political stability of the country but also more specific factors related to the management of natural resources, such as local impact and revenue management. Accordingly, the good management of the mining sector could help to enforce socio-political stability in producing countries. This statement does not claim a linear correlation between sustainable mining and the socio-political stability of a country. However, it suggests that sustainable mining can play a role in reducing such a risk.

To resume, although the current literature on phosphate criticality assessment explores the fragility of the supply due to social, environmental, and political conditions in

[1] Phosphate functionality: refers to three key intermediate phosphate products involved in the production of numerous final chemical products (see Chapter 4, section 4.1). The identified products are the beneficiated phosphate rock 29%, the phosphoric acid and the white phosphorus.

producing countries, it does not propose a complete image of the interaction between factors, which can lead to the disruption of the supply. According to Nedelciu et al. (2019), a clear and complete reporting on phosphate mining environmental-social externalities as well as international conflicts allows a better evaluation of global sustainability commitments to achieve global phosphate governance [142]. Therefore, it is crucial to understand the socio-political and environmental factors of phosphate rock production in a more specific local and national context.

8.3 The main outcomes of the Tunisian phosphate mining case study

In this section, two significant outcomes are discussed: first, the general concept of phosphorus criticality from the perspective of the producing country based on the experience of Tunisia. Second, the interrelationship between environmental and social impacts, based on the results of the E-LCA (Chapter 6) and the S-LCA (Chapter 7).

8.3.1 The criticality of phosphorus from the perspective of the producing country

The two dimensions of the mineral criticality are their supply risks and their importance to the national economy. The concept of criticality has been adopted from the perspective of countries that rely on imports, rather than on supplying countries. The concept of criticality, however, also applies to producing countries. As a matter of fact, the mineral could be critical to producing countries as the contribution of the mining sector to the national economy could be very high. According to the fourth edition of the Mineral Contribution Index (MCI[2]), the mineral sector remains the first driver of many low- and middle-income economies. The poorest countries remain dependent on mining despite a decline in commodity prices [110].

Besides the national economy's dependency on the mining sector, the mineral can be subject to supply risk due to internal conflict and local social instability. The case of phosphate rock mining in Tunisia is an excellent example to show how the phosphate sector

[2] MCI: Mining contribution Index is a composite of four 4 indicators: the export contribution, the evolution of export contribution, mining production value as percentage of GDP and mining production value minus the normal costs

(mining and fertilizer production), which is crucial for the health of the Tunisian economy, was heavily impacted by the social movement that started in the mining area.

Between 2010 and early 2011, the wave of socio-political revolutions, which started in the mining region, shocked the whole country (see section 0). Riots and civil resistance took place on the streets across the country, and faced police violence to reach the political power in position in January 2011 and led to a change in the system by ousting the president and its regime. The goals of the movement were employment, more social justice, reducing regional economic disparity, and health inequity between coastal regions and inland regions. Despite the wealth of the mining region and the billions of dinars generated by the phosphate sector, the mining region was for a long time the most marginalized area of the country in terms of social development, such as access to health and economic development, as the economic activity is highly concentrated. The rate of unemployment was 28% in 2010 in the mining area, twice the national unemployment rate.

As a result of the social movement, the phosphate sector suffered from an unprecedented economic crisis. The phosphate sector is one of the biggest industries for the Tunisian economy. It contributed directly to 2% to 4% of the Tunisian GDP before 2011 [15, 89], and it represented 7% of the total exportation in 2011 [139]. A single day of production shutdown in the CPG and the Tunisian Chemical Group (GCT) resulted in an average loss of 3 million dinars (nearly 1.25 million dollars) [12].

Moreover, the social instability that took the phosphate mining region in Tunisia hostage led to a drop in the production of phosphate rock from 8 MMT in 2010 to 2.2 MMT in 2011. The company is still unable to recover and regain its full production potential registered before the revolution [35]. The annual production in 2016 was 3.33 million tons and 3.94 in 2017 [53, 55]. The sales of beneficiated phosphate rock in 2016 and 2017 are down by minus 53% compared to 2010 [36, 38].

Consequently, the production of the fertilizer industry dropped 45% from 1.67 MMT in 2010 to 733 thousand metric tons in 2011 [68]. In 2011 the planned profit for the sector was planned to reach 1000 million dinars as the price of the commodity on the global market increased [68]. However, the sector made only 200 million dinars in 2011 compared to 825 million dinars in 2010 before the revolution [68]. The international market share of the Tunisian phosphate fertilizers shrank. Tunisia had lost 50% of its share in the Indian market, 50% of the Brazilian market, and lost the totality of the Turkish market [12]. The situation was exacerbated because of the increase of competition as new countries, such as Saudi Arabia and Peru, entered the market in 2012, and Morocco extended its production capacity.

The actions and measures to calm down the social anger, to stabilize the region, and to recover the sector had a focus of reducing the local unemployment. The CPG proceeded to hire workers, outsourced the transport of beneficiated phosphate rock and by-products to the company STTPM and other private contractors, and created the company of the environment, planting, and gardening of Gafsa in 2011. Furthermore, in 2017 the government ordered the reopening of the mining site of Meknassi and the creation of a beneficiation plant in the area of Sidi Bouzid to ease the tension between regions.

The described measures did not target the main factors of the problem; they treated the symptoms and delayed addressing the cause by sustaining the same development model. For instance, the number of workers at the CPG increased from 4,898 in 2010 to 6,619 in 2016. The department of the environment, planting, and gardening hired 3,100 individuals in 2011, and the number increased to reach 24,700 in 2016. However, the operating income declined dramatically. The deficit of the CPG was 109.8 million dinars in 2016. The deficit of the company of the environment was 336 thousand dinars as the personal expenses increased, and the company did not register any income since its creation (data from the Annual general ordinary meeting of the GCT, 2018) [90].

Furthermore, those solutions created other social and environmental problems. As the personal expenses exploded in the last several years and the productivity and profitability of the company was shrinking, the company could not afford jobs for everyone in the region. The local population is still unsatisfied with the hiring solution, and they believe that the hiring process is unfair. Consequently, more strikes and sit-ins were organized in front of the production sites of the company and blocked the transportation routes. According to the Tunisian social observatory (OST) (2018), 791 protest movements in 2018 occurred in the governorate of Gafsa. Most of the protest movements had socio-economic motivations: to demand employability, the improvement of social conditions, and the improvement of public services [80]. In the governorate of Sidi Bouzid – where a beneficiation was promised – 881 protests were organized to claim a clear answer regarding the future of the project, which has been on hold since 2016, demanding its implementation as soon as possible [80].

In addition to the social failure of those solutions to stabilize the region, they will worsen the environmental impacts of the phosphate production and threaten the livelihood of the region. For instance, by favoring transportation by truck, transport by rail dropped from 28.3 MMT of beneficiated phosphate rock in 2010 to 3.3 MMT in 2015 [35]. The solution did not take into consideration the environmental dimension as transportation with the truck is one of the highest emitters of GHGs in the lifecycle of the production of beneficiated phosphate rock. Sidi Bouzid is mainly an agricultural region; bringing phosphate activity to that region will have negative environmental impacts, particularly on the water sources, and therefore on the agricultural sector.

Those measures do not offer a sustainable development plan for the region by bringing investment opportunities in sectors that are more viable in the long term. The environmental issues, in particular, water source protection, are strategic for the stability of the region [125].

To summarize, due to social instability in the mining region, the production was heavily affected. The development of the short-term solution is inefficient and even has deepened the socio-environmental problems in the region. Phosphate rock is also critical for Tunisia as it is a barometer of the social situation [63]. It can maintain social stability, but in the case of imbalance, it triggers social instability with high economic consequences for the country.

8.3.2 The interrelations between environmental and social impacts

The results of the environmental and social impacts discussed in chapter 6 and chapter 7 show that the investigated environmental impacts have strong social dimensions. According to DI NOI and Ciroth (2018), for the assessment of the environmental and social impacts of mining, it is essential to take into consideration the interrelationship between both dimensions as the burden of one dimension can shift to another [47]. Moreover, the environmental and social impacts are not only primary impacts, which are immediately occurring in the mining site, but also secondary impacts related to infrastructure development and change in local economies [201]. These secondary impacts are a slow process, and they cause long-term changes. According to Macombe et al. (2013), in S-LCA the social impacts are the consequences of changes that entail a specific effect [126]. These effects could generate negative phenomena among an affected group of people (workers, local population, and society) [126].

Figure 8.1 represents the environmental and social change process, from the bottom of environmental impacts to socio-ecological effects to end up on social change, which are the long-term impacts such as high unemployment, change in health status, and very low economic diversification. Those social changes would have a high risk of triggering social conflicts in the region.

The social changes are grouped into three clusters:

- **Change in human health and safety**: due to the exposure of workers and local population to long-term hazardous elements emitted in the air, soil, and water.
- **Change in the livelihood of the region**: due to soil pollution and water depletion, the agricultural and pasture potential of the region is declining, reducing more and more the economic opportunities as the economic activity is already highly concentrated.
- **Change in the livability of the region**: due to air pollution with dust, the deterioration of water quality, and water interruption in households.

Besides the interdependence of environmental and social impacts in the mining region, potential negative economic effects may occur. Indeed, in the absence of social stability, the company is not able to achieve good economic profitability. The company struggles to reach its production potential due to strikes and sit-ins. Negative environmental impacts are the growing medium for social and economic injustice that can create a feeling of being neglected by the company and the responsible authorities (see Figure 8.1).

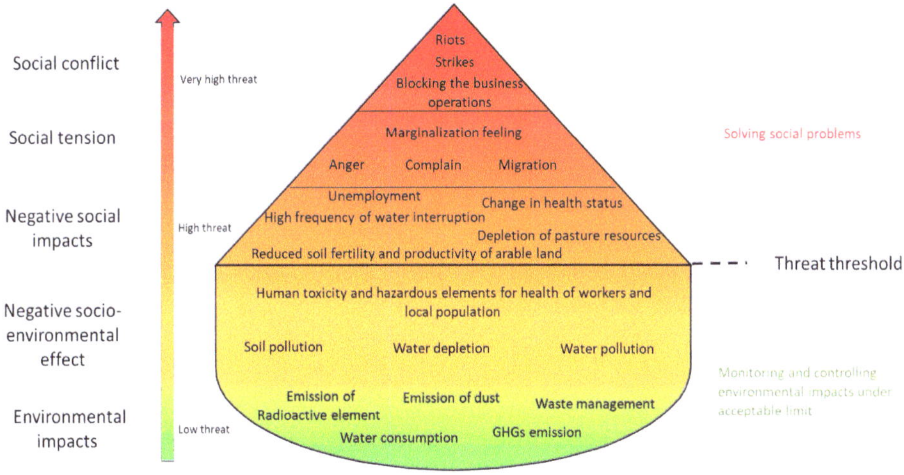

Figure 8.1: Environmental -social impact pathway from environmental effect to socio-environmental effect to social changes based on the results of E-LCA and S-LCA

The social and economic injustice is reducing the livelihood and livability of the region. They also reduce the attractiveness of the region to investments and economic activity diversification. The feeling of marginalization leads to social animosity towards the company (see chapter 7). As long as the situation is not kept under an acceptable risk-level, social anger can turn into riots, strikes, and attempts to block industry production.

Environmental and social impacts are inevitable for any mining project. Without the implementation of early-stage measures to maintain and control the negative impacts (very low to low risk), the threat of social change will lead to high costs for solving problems and reduce the profit of the company.

As the situation would reach the point of a negative social change, the company will allocate more effort and funding to solve external problems than focusing on increasing its profitability [100]. The company, as well as the governmental agencies with the participation of the local population, are better able to reduce secondary impacts and mitigate their changes by defining where environmental sensitivities exist [201].

Therefore, to promote better livability and livelihood of the region, the mining company CPG has to minimize, control, and offset the local ecological damages and ensures an equitable distribution of social services and economic investment. The management of environmental impacts and building the social responsibility capability of the company have to go hands in hands.

8.4 Measures toward global sustainable phosphorus management

This final section aims to close up the study with recommendations and measures toward a better global sustainable phosphorus management. The regional and national context of phosphate rock mining and production has to be integrated into the assessment of global phosphate supply.

8.4.1 Addressing the criticality of phosphorus from a global perspective

There is no standardized method to assess the criticality of minerals (see chapter 4). The method can change according to the scope of the assessment, which could be either national/regional or regarding the sector scope and the time frame [2]. In national reports, the criticality of a mineral is defined by its importance to the domestic economy and its vulnerability to external supply restriction. The critical minerals are those involved in mega sectors with a high contribution to the GDP of the country.

However, the assessment of phosphorus supply risks through its supply chain shows that the supply chains of minerals are complex and dispersed globally, as the mineral can be used in numerous industries. Thus, it is essential to study risk factors at an earlier stage of the supply chain.

Moreover, phosphorus is implicated not only in a sector with high economic importance, but also has a fundamental role in assuring food security of the growing global population. Hence, the definition of phosphorus criticality has to be aligned to the values of the Sustainable Development Goals (SDGs) and the targets of the concept of Planetary Boundaries [24, 170]. According to Scholz et al. (2014). phosphate management has to achieve the aspect of intra-and intergenerational justice, which are included in the Brundtland definition of sustainable development definitions [175]. Sustainable phosphorus management must encompass more justice in the global use of phosphorus and respect the planetary boundaries.

The high disparity of phosphorus usage for agriculture between different regions of the world (see Figure 8.2) and the overconsumption may induce the risk of socio-economic crisis in vulnerable areas as it deepens the inequalities towards access to food production and water. By addressing phosphorus criticality as a global problem, it helps to achieve some sustainable development goals such as reducing poverty, hunger, and improving access to clean water. Phosphorus criticality has to be a universal challenge to assure better, equitable, and sustainable phosphorus management.

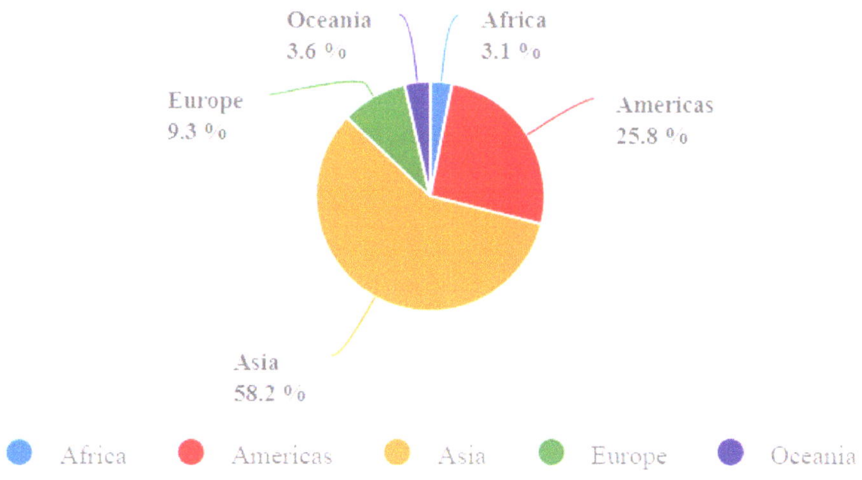

Figure 8.2: Nutrient phosphate P_2O_5 (total) agricultural use shares by region; Average 2002-2017 [79]

The criticality assessment of phosphorus needs to adopt a global perspective to achieve sustainable phosphorus management. It was suggested by Graedel et al. (2012) that "*It is from the global level perspective, and from a temporal horizon of half a century or more, that one might ask the question: Is society today using elements at a rate that permits the next generation to acquire them to the extent that might be needed?*" [87].

From this perspective, this study suggests that the definition of mineral criticality has to be more inclusive. A critical material should be defined as followed:

> "A critical material is a material that performs an essential and non-substitutable function, and it is subject to a risk of interruption or decrease of its supply, which will lead to negative economic, and or a social, and or environmental consequences for one or many actors in its supply chain."

8.4.2 The role of responsible mining to mitigate social instability

According to Hustrulid et al. (2013), "Responsible Mining" is the approach to establishing appropriate levels of environmental and social mitigation while achieving an acceptable economic result. The mining company must consider the environmental and social risks in an early stage when planning a new project as well as while the project unfolds [100]. Therefore, the environmental and social mitigation costs have to be included in the profit equation[3] to achieve an optimum between the economic responsibility, the environmental responsibility, and the social responsibility's components of responsible mining [100]. Moreover, according to the OECD, in high-risk areas, mining companies and those involved in the minerals supply chain can generate economic growth, sustain livelihoods, and foster local development [150].

The regional and national situation in producing countries has a high effect on the global supply. As phosphate rock is highly geo-concentrated, fostering more responsible mining practices in developing countries would help to reduce socio-economic and political instability. Consequently, mining in stable countries would improve the sustainable supply of these resources on the global market. Efforts have to be engaged globally to empowering emerging countries to reform their regulatory framework and to reinforce natural

[3] Profit equation: Profit/unit= Price/unit- (Extraction cost/unit+ Environmental cost/ unit+ Social cost/unit) [100]

resources governance as well as empowering societies and local communities by cultivating a culture of civil society and citizens who work to influence the political and social sphere positively.

It is better to foresee challenges in mining areas to avoid a supply risk than to let socio-political instability erupt. This is not possible without collaboration with mining countries to achieve responsible mining. This idea has already been discussed by the Critical Material Alliance (CRM Alliance): "Efforts need to be directed toward minimizing supply chain risks rather than continuing to develop methodologies and performing studies toward the identification of a CRM" [40].

Moreover, to mitigate the risk along the supply chain, it is important to identify early the challenges for emerging countries in which economies rely on mining. In most cases, emerging countries do not dispose of the tools and the know-how to establish policies and legal frameworks. Solutions such as economic cooperation, trade agreements, and technological support must be coupled with social-ecological empowerment.

Natural resources have always been in the center of all geopolitical competition. It is not possible that phosphate resources would be managed as a common good for the world's population to feed the world and assure justice for food production. It is a utopia to suggest to nations and governments to give up their sovereignty over the natural resource to a global common management strategy to eradicate hunger and poverty. The control of natural resources is one of the competitive advantages for economies.

However, working together to strengthen a global framework is possible through international institutions. A good example is the World Trade Organization (WTO), which came up after a long history of global trade wars and global instability to set the rules of international trade peacefully and to manage conflicts. It is not a utopia to aspire for international cooperation to organize mineral supply chains in a better way. The OECD Due Diligence Guidance for Responsible Supply Chains of Minerals is a big step towards uniting efforts of developed countries, emerging countries, and the industry to strengthen the transparency and sustainability of minerals supply chains [150].

8.4.3 Improving resource efficiency and increasing recycling

The phosphate mining sector is still dominated by a linear model that generates a significant loss of mineral, exerts high pressure on the current mineral resources and reserves, and creates high social and environmental impacts. However, the circular economy paradigm offers principles to close the loop of material flow through the reuse and recycling of materials along their supply chains to form "circular supply chains," reducing negative environmental and societal impacts as part of a sustainable system transition [83].

The central axis of actions to improve the sustainable management of phosphorus and create a circular phosphorus supply chain are: (1) improving the efficiency of the mining operations and the beneficiation process; (2) increasing the resources efficiency use through the supply chain; and (3) closing the loop by recycling and recovery of phosphorus from waste food, municipal and industrial wastewater.

Losses of phosphorus are at all stages of the value chain, from mining and fertilizer production to food processing and food consumption. It is estimated that losses during mining using open-pit extraction are around 5 to 50% of P_2O_5, whlie losses during the beneficiation in North Africa is around 30% [215]. According to the results of the inventory analysis of phosphate mining in Tunisia (see Chapter 6), 20% of crude phosphate rock is lost during crushing and screening operations, and 20 to 25% of P_2O_5 is lost during beneficiation.

During the application phase in agriculture, phosphorus is washed off the land and goes into aquatic environments where it causes eutrophication[4]. It is estimated that the current agri-food system is inefficient in phosphorus use; only one-fifth of the phosphorus mined for food production finds its way into food consumed by the global population each year [39].

Assessing the phosphorus flow and balance in the European Union, Van Dijk et al. (2016) found out that points of high phosphorus emissions in the food chain in Europe are the household's food consumption and the food processing industry [214].

The recovery of phosphorus from municipal and industrial wastewater may create a strategic opportunity for European countries to reduce their dependency on the importation of primary phosphate and to improve their resource efficiency and environmental performance. According to the European Sustainable Phosphorus Platform, recycled phosphorus products could replace 20-40% of the EU mineral phosphate fertilizer [99]. However, investing in phosphorus recovery from wastewater needs to be included in a comprehensive approach to improving the sustainable use of phosphate rock sources. Scenarios

[4] Eutrophication happens due to the overabundance of phosphorus or nitrate in the aquatic environment, algae use these nutriments to grow and reproduce faster as it has to be, the overproduction and dense growth of algae causes the depletion of dissolved oxygen and creating by consequence an anaerobic environment 'dead zones' and fish die in lakes, rivers and oceans

based on the environmental, social, and economic performance of both primary and secondary phosphorus products would allow the development of the strategy to define the optimal combination of primary and secondary phosphate fertilizers.

9 General Conclusion

Phosphorus criticality has a humanitarian dimension, as phosphorus serves a basic human need, which is food production to feed the global population. Phosphorus criticality has not to be addressed only as a local problem but also as a global problem. Moreover, phosphorus criticality has to be more aligned with sustainable development goals to reduce poverty in the world, to eradicate hunger, and to reduce inequalities.

The keys findings of this study are the following:

- Phosphorus is not scarce in the short term as the current mined reserve stock to feed the world can cover 300 years at the current consumption rate (the defined resources can cover more than 1,000 years at the current consumption rate). Also, the reserves are dynamic because of technology. However, the quality is decreasing, and the environmental impacts related to P mining are increasing. Additionally, the global consumption rate of phosphorus is high and continues to increase. The action toward sustainable phosphorus management also needs to target more responsible phosphorus consumption.
- The supply risk factors are the social and environmental impacts of phosphate rock mining in producing countries. Environmental and social impacts are correlated and together can trigger negative economic impacts. The environmental burden of mining can impact the health and wellbeing of the local population negatively. The social change in the mining area also can induce negative environmental impacts if the used technologies and the proposed strategy led to creating more jobs at the expense of the environment.
- Phosphate rock is also critical for producing countries, as the health of their economy depends on the performance of the mining sector. The local population is a powerful stakeholder; the mining sector needs to manage the local population carefully to avoid social anger. The socio-environmental factors can trigger a shutdown of the mining sector and create an economic crisis.
- Responsible mining is a cornerstone of mitigating the supply risk factors and ensuring a sustainable supply chain. Good phosphate mining practices can support local social stability.

This study can be applied to other critical minerals to achieve sustainable supply chains, especially critical materials mined in developing countries with a high risk of environmental and social impacts. For instance, the mining sector is the first link of the supply chain, and its socio-environmental impacts in emerging economies are an early indicator of the supply disruption. Furthermore, the results of the environmental and social assessment of phosphate rock production can be used in a broader scope to study scenarios and strategies for importing countries to combine primary and secondary phosphorus resources to produce a more sustainable fertilization strategy.

A take-home message of this work is that sustainable phosphorus management is a global issue. Therefore, to mitigate phosphorus criticality, it needs a global action: first the implementation of responsible mining in emerging countries; second better resources efficiency during the use phase (mainly in agriculture and food consumption); and finally, reducing the loss of phosphorus into the environment along its value chain.

Bibliography

[1] Acero, A. P., Rodriguez, C., and Ciroth, A. 2015. LCIA methods: Impact assessment methods in Life Cycle Assessment and their impact categories. 1.5.4.

[2] Achzet, B. and Helbig, C. 2013. How to evaluate raw material supply risks—an overview. Resources Policy 38, 4, 435–447.

[3] African Manger. 2019. Le chantage de Bouzaïene, qui ruine la CPG, continue impunément. African Manger 2019 (Sep. 2019).

[4] Agricultural Research for Development. 2016. Social LCA research book: Social evaluation of the life cycle application to the agriculture and agri-food sectors. Observatoire des Marchés du CIRAD, Sète, France.

[5] Airbus, L./ C. and Maxar technologies, M. d. 2019. Current mining sites coverage 2017-2018. Google Maps.

[6] Atlas Copco Drilling Solutions LLC. 2015. DM45-series blasthole drills.

[7] Auvray, F. and Auvray, F. 2013. Les gisements des phosphates de Gafsa, Tunisie. phosphates deposits of Gafsa, Tunisia. Accessed 24 June 2020.

[8] Ayeb, H. 2011. Social and political geography of the Tunisian revolution: the alfa grass revolution. Review of African Political Economy 38, 129, 467–479.

[9] Ayeb, H. and Fautras, M. 2013. Pauvreté et Révolution en Tunisie / Poverty and Revolution in Tunisia Cartes / maps + Article « Révolution de l’Alfa » (Alfa Grass Revolution). https://habibayeb.wordpress.com/2013/05/15/pauvrete-et-revolution-en-tunisie-poverty-and-revolution-in-tunisia-cartes-maps/. Accessed 20 August 2020.

[10] Bard, D. and Ouertani, A. 2017. Etéude d'impact de la pollution industrielle sur la santé de l'Homme à Gabès. Study of the impact of industrial pollution on the human health in Gabes 2016/372830/2.

[11] Bates, B., Kundzewicz, Z. Z.W., and Palutikof, J.P., Eds. 2008. Climate change and water. Technical Paper of the Intergovernmental Panel on Climate Change 6. IPCC Secretariat, Geneva.

[12] Ben Ahmed, L. 2018. Tunisie – Crise du bassin minier : le gouvernement opte pour la fermeté. https://www.aa.com.tr/fr/afrique/tunisie-crise-du-bassin-minier-le-gouvernement-opte-pour-la-fermet%C3%A9-/1079625. Accessed 20 May 2020.

[13] Bio by Deloitte. 2015. Study on Data for a Raw Material System Analysis. Roadmap and Test of the fully operational MSA for raw materials final report.

[14] Bonnet, C. 2011. Les Phosphates. Phosphate. https://www.la-sim.org/images/doc_gratuite/les-phosphates.pdf. Accessed 19 June 2020.

[15] Boubaker, S. and Hassen, M. 2016. La Compagnie des Phosphates de Gafsa (CPG) : État des lieux de la gouvernance et recommandations.

[16] Bouchrika, R. 2017. Mesures sismiques et éude des effets de tirs dans les mines à ciel ouvert. Sismic measurement and study of the blasting effects in open pit mine, Metlaoui, Tunisia.

[17] Boujlel, H., Daldoul, G., Tlil, H., Souissi, R., Chebbi, N., Fattah, N., and Souissi, F. 2019. The Beneficiation Processes of Low-Grade Sedimentary Phosphates of Tozeur-Nefta Deposit (Gafsa-Metlaoui Basin: South of Tunisia). Minerals 9, 1, 2.

[18] Brahmi, M., Zouari, S., and Rossi, M. 2014. L'industrie minière et ses effets écologiques. État socio-économique et environnemental dans le bassin minier tunisien. edyte 17, 1, 109–120.

[19] British Geological Survey, Bureau de Recherches Géologiques et Minières, Deloitte Sustainability, Directorate-General for Internal Market, Industry, Entrepreneurship and SMEs, and TNO. 2017. Study on the review of the list of critical raw materials. Critical raw materials factsheets. Publications Office of the European Union, Luxembourg.

[20] Britt, A., Summerfield, D., Senior, A. 2016. Australia's Identified Mineral Resources 2016. Geoscience Australia.

[21] Bronzo Perasso- carrière et bétons. 2018. Demande d'autorisation environnementale unique pour l'exploitation d'une installation classe pour la protection de l'environnement, Renouvellement de l'autorisation d'exploiter avec extension de la carrière de sainte-Marthe: Pièce 5- Demande d'autorisation les effets sur la santé. Application

for a single environmental permit for the operation of an installation classed for environmental protection, Renewal of the authorization to operate with extension of the quarry of Sainte-Marthe: Exhibit 5- Application for authorization health effects. Accessed 22 February 2020.

[22] Bureau International du Travail, Service des principes et droits fondamentaux au travail, and Institut National de la Statistique. 2017. Enquête Nationale sur le travail des enfants en Tunisie,2017. National survey on the child labor in Tunisia,2017.

[23] Burreau, B., Hossie, G., and Lutfalla, S. 2013. Approvisionnements en métaux critiques, un enjeu pour la compétitivité des industries française et européenne. Document de travail 2013-04.

[24] Carpenter, S. R. and Bennett, E. M. 2011. Reconsideration of the planetary boundary for phosphorus. Environ. Res. Lett. 6, 1, 14009.

[25] Caterpillar, I. 2010. Specalog for D10T Track-Type Tractor, AEHQ6029.

[26] Caterpillar, I. 2011. Specalog for 789C Mining Truck AEHQ6167.

[27] Caterpillar, I. 2012. Specalog for Cat 789D Mining Truck AEHQ6237-03.

[28] CHARNI MAJOUBI, H., ABBES, A., ABOUDI, A., KHAYATI, S., GRAUBY, A., GHARBI, H. A., and MTIMET, S. 1991. Étude de la radioactivité naturelle dans le sol du sud tunisien Région de Gafsa Tozeur. Radioprotection 26, 3, 537–549.

[29] Ciroth, A. and Eisfeldt, F. May/2016. PSILCA- A product social impact life cycle assessment database. Version 1.1.

[30] CNES/Airbus. 2020. Research Center, Metlaoui. Google Maps.

[31] CNES/Airbus and Maxar technologies, M. d. 2020. Dépot Maintenance Metlaoui (DMM) CPG. Google Maps.

[32] CNES/Airbus and Maxar technologies, M. d. 2020. Laverie Kef Eddour. Google Maps.

[33] CNES/Airbus and Maxar technologies, M. d. 2020. Laverie Moulares CPG. Google Maps.

[34] 2018. Code Minier. In Loi n° 2003-30 du 28 avril 2003 portant promulgation du Code Minier.

[35] Companie des phosphates de Gafsa CPG. 2016. Rapport annuel 2015.

[36] Companie des phosphates de Gafsa CPG. 2018. Rapport d'activité générale pour l'année2016. Document interne.

[37] Companie des phosphates de Gafsa CPG. 2018. Compagnie des Phosphates de Gafsa: Historique. Company of phosphates of Gafsa: history. https://www.cpg.com.tn/. Accessed 24 June 2020.

[38] Companie des phosphates de Gafsa CPG. 2019. Rapport d'activité générale pour l'année 2017. Document interne.

[39] Cordell, D. 2010. The Story of Phosphorus. Sustainability implications of global phosphorus scarcity for food security. Dissertation, Linköping University faculty of arts and sciences.

[40] CRM Alliance. 2019. CRM list and methodology. http://criticalrawmaterials.org/policy/crm-policy/.

[41] De Ridder, M., Jong, S. de, Polchar, P., and Lingemann, S. 2012. Risks and opportunities in the global phosphate rock market. Robust strategies in times of uncertainty. Rapport / Centre for Strategic Studies no. 17 | 12 | 12. The Hague Centre for Strategic Studies (HCSS), Den Haag.

[42] DERA - Deutsche Rohstoffagentur in der Bundesanstalt für Geowissenschaften und Rohstoffe. 2012. DERA-Rohstoffliste 2012. Angebotskonzentration bei Metallen und Industriemineralen - Potenzielle Preis- und Lieferrisiken. DERA Rohstoffinformationen 10, Berlin.

[43] DERA - Deutsche Rohstoffagentur in der Bundesanstalt für Geowissenschaften und Rohstoffe. 2014. DERA-Rohstoffliste 2014. Angebotskonzentration bei Metallen und Industriemineralen - Potenzielle Preis- und Lieferrisiken. DERA Rohstoffinformationen 24.

[44] DERA - Deutsche Rohstoffagentur in der Bundesanstalt für Geowissenschaften und Rohstoffe. 2016. DERA-Rohstoffliste 2016. Angebotskonzentration bei mineralischen Rohstoffen und Zwischenprodukten – potenzielle Preis- und Lieferrisiken. DERA Rohstoffinformationen 32, Berlin.

[45] DERA - Deutsche Rohstoffagentur in der Bundesanstalt für Geowissenschaften und Rohstoffe. 2019. DERA Rohstoffliste 2019. Angebotskonzentration bei Metallen und

Industriemineralen - Potenzielle Preis- und Lieferrisiken. DERA-Rohstoffinformationen 40, Berlin.

[46] Dhibi, A., Elbarcharoui, M., and Fattah, N. 2017. Guided visit to the mining site Table Metlaoui, Washing plant n°4 and Research Center. Accessed 27 June 2020.

[47] Di NOI, C. and Ciroth, A. 2018. The importance of Three-dimension approach in LCA. A screening study on mining addressing environmental, social and cost aspects. LCA XVIII Conference.

[48] Di NOI, C., Eisfeldt, F., Ciroth, A., and Bizarro, D. 10-12 Decembre 2018. Complementarity of social and environmental indicators in assessing the sustainability of mining industry, Pescara (Italy).

[49] Direction de sécurité et de l'environnement. 2016. Rapport d'audit interne. Document interne.

[50] Direction de sécurité et de l'environnement. 2017. Tableau des indicateurs sur les accidents pour l'année 2016. Document interne.

[51] Direction de sécurité et de l'environnement. 2018. Statistique des acccidents et maladies professionnelles pour l'année 2017. Document interne.

[52] Direction générale de l'aménagement du territoire. 2011. Atlas du gouvernorat de Gafsa.

[53] Direction of Accounting. 2017. Tableau de bord de gestion et des indicateurs de performance de 2016: les coût complets. Internal document, 6.

[54] Direction of Accounting. 2017. Total consumption CPG. Internal document, 10. Accessed 29 June 2020.

[55] Direction of Accounting. 2018. Tableau de bord de gestion. Internal document, 7.

[56] Direction of human resources. 2018. Rapport de classification des employés selon la catégorie, le genre, le salaire, congès et séance de travail effectué. Document interne.

[57] Direction of human resources. 2018. Tableau de distribution des fonctions selon le genre. Document interne.

[58] Direction of mining planification. 2004. Synoptique de la préparation mécanique secteur Metaloui Kef Shfaier, Redeyf, Moulares. Intenal document. Accessed 27 June 2020.

[59] Direction of research and innovation. 2010. Etude de la qualité du phosphate de la Table de Metlaoui. Internal file, 4.

[60] Direction of the sector Metlaoui-Kef Schfaier. 2017. Activity report of the mining site Table de Metlaoui. Internal document, 9.

[61] Direction of the sector Metlaoui-Kef Schfaier. 2017. Bilan matière de la laverie 4, 12.

[62] Direction of the sector Metlaoui-Kef Schfaier. 2017. Exploitation planning and investment budget for 2018. Internal document, 8.

[63] Dumas, L. R. 2018. L'industrie des phosphates en Tunisie: un baromètre de la situation sociale. https://www.francetvinfo.fr/monde/afrique/economie-africaine/l-industrie-des-phosphates-en-tunisie-un-barometre-de-la-situation-sociale_3053945.html. Accessed 26 May 2020.

[64] Eisfeldt, F. Soca v.1 add-on – Adding social impact infor-mation to ecoinvent: Description of methodology to map social impact information from PSILCA v.1 to ecoinvent v. 3.3.

[65] EPA. Phosphate Rock Processing Background Report. In AP-42: Compilation of Air pollution, Vol. 1.

[66] EPA. Phosphate Rock Processing final section. In AP-42: Compilation of Air pollution, Vol. 1.

[67] Erdmann, L., Behrendt, S., and Feil, M. 2011. Kritische Rohstoffe für Deutschland. Identifikation aus Sicht deutscher Unternehmen wirtschaftlich bedeutsamer mineralischer Rohstoffe, deren Versorgungslage sich mittel- bis langfristig als kritisch erweisen könnte", Berlin.

[68] Espace Manager. 2011. Tunisie: les recettes des phosphates fondent en 2011. https://www.espacemanager.com/tunisie-les-recettes-des-phosphates-fondent-en-2011.html. Accessed 22 May 2020.

[69] EU Ad-Hoc Working Group on Raw Materials. 2014. Report on critical raw materials for the EU.

[70] European Commission. 2010. Critical raw materials for the EU. Report of the Ad-hoc Working Group on defining critical raw materials Ref. Ares(2014)2187691 - 02/07/2014. European Commission.

[71] European Commission. 2011. Communication from the Commission to the European Parliament, the Council, the Eropean Economic and Social Committee of the regions: Tackling the challenges in commodity markets and on raw materials.

[72] European Commission. 2017. Methodology for establishing the EU list of critical raw materials. guidelines, Brussels. DOI=10.2873/769526.

[73] European Commission. 2017. Report from the commission to the European parliament, the council, the european economic an social committee and the committee of the regions on the implementation of the circular economy plan, Brussels.

[74] European Commission, Joint Research Center, and Institute for Environment and Sustainaility. 2010. International Reference Life Cycle Data System (ILCD) Handbook - General guide on LCA - Detailed guidance EUR24708, Luxembourg.

[75] European Environmental Agency. 2019. 2.A.5.a Quarrying and mining calculation model 2019 - Spreadsheet — European Environment Agency. https://www.eea.europa.eu/publications/emep-eea-guidebook-2019/part-b-sectoral-guidance-chapters/2-industrial-processes/2-a-mineral-products/2-a-5-a-quarrying-1/view. Accessed 7 March 2020.

[76] European Environmental Agency (EEA), Ed. 2019. 2.A.5.a Quarrying and mining of minerals other than coal. Technical guidance to prepare national emission inventories. EMEP/EEA air pollution emission inventory guidbook 2019.

[77] FAO. 2017. World fertilizer trends and outlook to 2020.

[78] Fattah, N. 2005. Geological structure of phosphorites in the Gafsa basin (Tunisia) and their geochemical specialization. Dissertation, The Russian State Geological Prospecting University (MGRI-RSGPU).

[79] Fodd and Agriculture Organization of the United Nations. 2019. FAOSTAT: Fertilizers by Nutrient:. http://www.fao.org/faostat/en/. Accessed 23 August 2020.

[80] Forum Tunisien des droits économiques et sociaux. 2018. Rapport du mois de décembre 2018 des mouvements sociaux, suicides et violences. Report of december 2018 on the social mouvement, suicide and violence 64. Observatoir social tunisien.

[81] Frischknecht, R. 1998. Life cycle inventory analysis for decision-making.

[82] Gantner, O. 2016. Ressourcenstrategische Betrachtung der Kritikalität von Phosphor. Dissertation, Shaker Verlag GmbH.

[83] Geissler, B., Hermann, L., Mew, M., and Steiner, G. 2018. Striving Toward a Circular Economy for Phosphorus: The Role of Phosphate Rock Mining. Minerals 8, 9, 395.

[84] Giurco, D., Mohr, S., Mudd, G., Mason, L., and Prior, T. 2012. Resource Criticality and Commodity Production Projections. Resources 1, 1, 23–33.

[85] Gobe, E. 2011. The Gafsa Mining Basin between Riots and a Social. HAL, halshs-00557826ff.

[86] Goedkoop, M., Heijungs, R., Huijbregts, M., Schryver, A. de, Struijs, J., and Van Zelm, R. 2009. ReCiPe 2008 : A life cycle impact assessment method which comprises harmonised category indicators at midpoint and the endpoint level. Report 1: Characterisation.

[87] Graedel, T. E., Barr, R., Chandler, C., Chase, T., Choi, J., Christoffersen, L., Friedlander, E., Henly, C., Jun, C., Nassar, N. T., Schechner, D., Warren, S., Yang, M.-Y., and Zhu, C. 2012. Methodology of metal criticality determination. Environmental science & technology 46, 2, 1063–1070.

[88] Greendelta. December 19th, 2019. Openlca. GreenDelta.

[89] Groupe Chimique de Tunisie GCT. 2011. The Tunisian Phosphate sector. http://www.gct.com.tn/produits-services/secteur-des-phosphates/. Accessed 22 May 2020.

[90] Groupe Chimique de Tunisie GCT. 2018. The Annual general ordinary meeting of the GCT.

[91] Grunwald, A. 2010. Technikfolgenabschätzung. Eine Einführung. Gesellschaft, Technik, Umwelt Neue Folge, 1. Edition Sigma, Berlin.

[92] Hajem, S. 2015. Système tunisien d'information sur les causes de décès: Entraves spécifiques, synthèse des principaux résultats de l'année 2013 et perspectives. Tunisian information system on the causes of death: Specific obstacles, summary of the main results for 2013 and prospects. l'Unité de recherche et d'information sur les causes de vieillissement et les cause médicales de décès.

[93] Haldar, S. K. 2018. Mineral exploration. Principles and applications. Elsevier, Amsterdam, Netherlands.

[94] Hamed, Y., Ahmadi, R., Demdoum, A., Bouri, S., Gargouri, I., Ben Dhia, H., Al-Gamal, S., Laouar, R., and Choura, A. 2014. Use of geochemical, isotopic, and age tracer data to develop models of groundwater flow: A case study of Gafsa mining basin-Southern Tunisia. Journal of African Earth Sciences 100, 418–436.

[95] Hassen Mzoughi. 2017. Sami Tahri: " Un député possède 100 camions pour le transport le phosphate". Sami Tahri: " A deputy has 100 truck for the transport of phosphate. Kapitalis 2017 (Nov. 2017).

[96] Hatayama, H. and Tahara, K. 2015. Criticality Assessment of Metals for Japan's Resource Strategy. Mater. Trans. 56, 2, 229–235.

[97] Heni, W., Mokni, N., and Shafaie, A. 2018. Code minier de la tunisie. l'amélioration des aspects liés à la transparence et à la gouvernance.

[98] Heni, W. and Shafaie, A. 2017. Tunisia's Draft Law on Parliamentary Approval of Oil Contracts: Missed Transparency Opportunity? https://resourcegovernance.org/blog/tunisia%E2%80%99s-draft-law-parliamentary-approval-oil-contracts-missed-transparency-opportunity. Accessed 12 October 2019.

[99] Hermann, L. and Thornton, C. Developments in regulatory and downstream user perspectives in sewage sludge management in Europe: drivers, challenges and opportunities. In ECSM'2019: European Conference on Sludge Management.

[100] Hustrulid, W. A., Kuchta, M., and Martin, R. K. 2013. Open Pit Mine Planning and Design, 1. Fundamentals Volume 1, London.

[101] IHS Markit. 2017. Phosphorus and Phosphorus Chemicals. CHemical Economic Handbook. https://ihsmarkit.com/products/phosphorus-chemical-economics-handbook.html.

[102] INS. 2010. Enquete nationale sur la population et l'emploi 2010.

[103] INS. 2011. Enquête nationale sur la population et les ménages 2009.

[104] INS. 2014. Gafsa à travers le recensement général de la population et de l'habitat 2014.

[105] INS. 2015. Rapport annuel sur les indicateurs d'infrastructure-. Le comité général du développent sectoriel et régional.

[106] INS. 2016. Consommation et niveau de vie. consumption and standard of living. Flash consommation et niveau de vie 1.

[107] INS. 2016. Enquête Emploi et salaires auprès des Entreprise en 2014.

[108] INS. 2017. Indicateurs de l'emploi et du chômage: deuxième trimestre 2017. Employment and unemployment indicators: second trimester 2017 P0201. Institut National de la Statistique (INS).

[109] Institut National de la Statistique. 2019. Les comptes de la nation: Agrégat et tableaux d'ensemble 2013-2017, Méthodology et principaux résultats. Institut National de la Statistique (INS).

[110] International Council on Mining and Metal. 2018. Role of mining in national economies. Mining Contribution Index.

[111] International Fertilizer Association. 2019. Production and International Trade, World Phosphoric acid production by region. https://www.ifastat.org/supply/Phosphate%20Products/Processed%20Phosphates. Accessed 26 August 2020.

[112] International Fertilizer Association (IFA), Ed. 2018. Fertilizer Outlook 2018- 2022. Medium- Term Outlook for Fertilizer and Raw Materials Global Supply :2017/18-2022/23.

[113] International Monetary Fund. 2019. IMF Data access to macroeconomic and financial data : Government Finance statistics (GFS). https://data.imf.org/?sk=89418059-d5c0-4330-8c41-dbc2d8f90f46&sId=1437430552197. Accessed 12 March 2019.

[114] International Organization for Standardization. 2006. Environmental management — Life cycle assessment — Requirements and guidelines, Geneva Switzerland, ISO 14044:2006(E).

[115] International Organization for Standardization. 2019. The ISO Survey of Management System Standard Certifications . - Full results. https://isotc.iso.org/livelink/livelink?func=ll&objId=18808772&objAction=browse&viewType=1. Accessed 28 November 2019.

[116] International Trade Centre. 2019. Trade Map - List of exporters for the selected product (Natural calcium phosphates and natural aluminium calcium phosphates, natural and phosphatic ...). https://www.trademap.org/tradestat/Country_SelProduct_TS.aspx?nvpm=1%7c%7c%7c%7c%7c2510%7c%7c%7c4%7c1%7c1%7c2%7c2%7c1%7c2%7c2%7c1%7c1. Accessed 19 February 2019.

[117] Kasprak, A. 2016. The Desert Rock That Feeds the World. https://www.theatlantic.com/science/archive/2016/11/the-desert-rock-that-feeds-the-world/508853/. Accessed 21 August 2020.

[118] Kaufmann, D., Kraay, A., and Mastruzzi, M. 2010. The worldwide Governance Indicators. Methodology and Analytical issues. Policy research working paper 5430. The World Bank Development Research Group.

[119] Kellenberger, D., Althaus, H.-J., Jungbluth, N., Künniger, T., and Thalmann, P. 2007. Life Cycle Inventories of Building Products. Final report. Ecoinvent Data V2.0 No 7. Swiss Center for Life Cycle Inventories, Dübendorf, Swizerland.

[120] Khelifi, F., Besser, H., Ayadi, Y., Liu, G., Yousaf, B., Harabi, S., Bedoui, S., Zighmi, K., and Hamed, Y. 2019. Evaluation of potentially toxic elements' (PTEs) vertical distribution in sediments of Gafsa–Metlaoui mining basin (Southwestern Tunisia) using geochemical and multivariate statistical analysis approaches. Environ Earth Sci 78, 2.

[121] Khelifi, L. 2012. Contribution à l'étude géochimique des phosphates du bassin de Gafsa-Métlaoui : Exemple du Gisement d'Oum EL Khecheb. Master, FACULTE DES SCIENCES DE TUNIS.

[122] KOMATSU Mining Germany. 2005. KOMATSU PC3000.

[123] KOMATSU Mining Germany. 2013. Komatsu Prospekte PC5500.

[124] Kroop, S., Kaufhold, T., Lohmeyer, R., Mocker, M., Franke, M., Kranet, M., Böhme, L., Genslein, M., and Clauß, D. 2014. Analyse kritischer Rohstoffe für die Landesstrategie Baden-Württemberg ZO3 R12003-5. Fraunhofer-Institut für Umwelt,Sicherheits und Energietechnik-UMSICHT/ Universität Stuttgart-Insitut für Siedlungswasserbau, Wassergüte und Abfallwirtschaft, Baden Württemberg.

[125] Laabidi, W. 2017. Repenser le phosphate… Repenser l'environnement : Athimar. https://www.athimar.org/articles/details/repenser-le-phosphate-repenser-lenvironnement. Accessed 20 May 2020.

[126] Macombe, C., Leskinen, P., Feschet, P., and Antikainen, R. 2013. Social life cycle assessment of biodiesel production at three levels: a literature review and development needs. Journal of Cleaner Production 52, 205–216.

[127] Mancini, L. and Sala, S. 2018. Social impact assessment in the mining sector: Review and comparison of indicators frameworks. Resources Policy 57, 98–111.

[128] Mancini L., Eynard U., Eisfeldt F., Ciroth A., Blengini G., Pennington D. 2018. Social assessment of raw materials supply chains. A life-cycle-based analysis. JRC Technical Reports EUR 29632 EN. European Commission, Luxembourg. DOI=10.2760/470881.

[129] Maria Chiara Cavalleri, Alice Eliet, Peter McAdam, Filippos Petroulakis, Ana Soares, and Isabel Vansteenkiste. Concentration, market power and dynamism in the euro area.

[130] Marscheider-Weidemann, F., Langkau, S., Hummen, T., Erdmann, L., Tercero Espinoza, L. A., Angerer, G., Marwede, M., and Benecke, S. 2016. Rohstoffe für Zukunftstechnologien 2016. Auftragsstudie. DERA Rohstoffinformationen 28. Deutsche Rohstoffagentur (DERA) in der Bundesanstalt für Geowissenschaften und Rohstoffe (BGR), Berlin.

[131] Metso Corporation. 2018. MHC 250 Standard Equipment Data Sheet 3662-01-18-ENG-MNG.

[132] Michaud, D. 2016. Rotary Dryer Design & Working Principle. https://www.911metallurgist.com/blog/rotary-dryer-design-working-principle. Accessed 19 April 2020.

[133] Mikkola, H., Ed. 2018. Seabirds. InTech.

[134] Miller, T., Kim, A. B., and Roberts, J. M. 2019. 2019 index of Economic Freedom. The Heritage Foundation.

[135] Mine-Engineer.com. 2020. Basics of an open pit mine. http://www.mine-engineer.com/mining/open_pit.htm. Accessed 7 March 2020.

[136] Mining planification direction. 2018. Measured and demonstrated reserves in Gafsa mining region. Internal document, 2. Accessed 27 June 2020.

[137] Ministère des affaires locales et de l'environnement. 2018. Arrêté du ministre des affaires locales et de l'environnement et du ministre de l'industrie et des petites et moyennes entreprises du 26 mars 2018, fixant les valeurs limites des rejets d'effluents dans le milieu récepteur. In Journal Officiel de la République Tunisienne, 823–838.

[138] Mokaddem, N. 2017. Etude des aquifères et gestion des ressources en eau dans le bassin de Gafsa Nord. Doctoral, Université de Sfax.

[139] Mtimet, N., Loridan, M., and Rajaonarivelo. 2014. Tunisie: Perspectives des entreprises. Série de l'ITC sur les mesures non tarifaires MAR-14-250.F. Ecole spérieure d'agriculture de Mograne, International Trade Center (ITC).

[140] Muthu, S. S. 2019. Social Life Cycle Assessment. Case Studies from Agri and Food Sectors. Environmental Footprints and Eco-design of Products and Processes. Springer Nature Singapore Pte Ltd., Singapore.

[141] National Research Council. 2008. Minerals, critical minerals, and the U.S. economy. National Academies Press, Washington D.C.

[142] Nedelciu, C.-E., Ragnarsdóttir, K. V., Stjernquist, I., and Schellens, M. K. 2019. Opening access to the black box: The need for reporting on the global phosphorus supply chain. Ambio.

[143] Nicolas, F. 2016. Compte rendu d'une enquête abrégée à Redeyef (Sud-Ouest Tunisia). Accessed 24 June 2020.

[144] NRGI. 2017. Natural resource governance Index 2017.

[145] Numbeo.com. 2019. Rankings by Country of Average Monthly Net Salary (After Tax) (Salaries And Financing). https://www.numbeo.com/cost-of-living/country_price_rankings?itemId=105&displayCurrency=TND. Accessed 19 July 2019.

[146] Office de développement du sud. july/2017. Gouvernorat de Gafsa en chiffre 2016. Ministère du développement, de l'investissement et de la coopération internationale, la République Tunisienne.

[147] Office de Développement du Sud. 2017. Gouvernorat de Gafsa en chiffre 2016. Ministère du développement,de l'investissement et de la coopération internationale.

[148] Office National des Mines. 2020. Industrie minérale extractive en Tunisia. http://www.onm.nat.tn/fr/index.php?p=indminier. Accessed 16 September 2020.

[149] ONEQ. 2013. Rapport annuel sur le marché du Travail en Tunisie.

[150] Organisation for Economic Co-operation and Development. 2013. OECD Due Diligence Guidance for Responsible Supply Chains of Minerals from Conflict-Affected and High-Risk Areas. OECD Publishing.

[151] Organisation for Economic Co-operation and Development. 2015. Tunisia: A Reform Agenda to support Competitiveness and Inclusive Growth.

[152] Organisation for Economic Co-operation and Development. 2019. Gender pay gaps for full and part-time workers. Database, LMF5. http://www.oecd.org/els/family/database.htm. Accessed 12 March 2019.

[153] Organisation for Economic Co-operation and Development. 2019. Unemployment rate - OECD Data. https://data.oecd.org/unemp/unemployment-rate.htm. Accessed 22 June 2020.

[154] Organisme Professionnel de Prévention du Bâtiment et des Travaux Public. 2003. Les principaux explosifs utilisés dans les travaux publics. the main explosives used in public works. Accessed 22 February 2020.

[155] Paie-tunisie.com. Paie Tunisie -Smig et Smag. Tunisia Salary: Interprofessional guaranteed minimum wage SMIG and Agriculture guaranteed minimum wage SMAG. https://www.paie-tunisie.com/369/fr/smig-et-smag.aspx. Accessed 12 March 2019.

[156] Picard, r. and Guidara, A. 2016. Municipalité de Gafsa: Rapport final sur la performance de la gestion des finances publiques. Municipality of Gafsa: Final report on the performance of public financal management.

[157] polymerdatabase.com. 2020. Properties of Polyacrylamides. https://polymerdatabase.com/polymer%20classes/Polyacrylamide%20type.html. Accessed 3 December 2020.

[158] PRé Sustainability. 2020. ReCiPe. https://www.pre-sustainability.com/recipe. Accessed 4 September 2020.

[159] Primas, A., Capello, C., and Hischier, R. 2007. Phosphate rock. In Life Cycle Inventories of Chemicals, Swiss Center for Life Cycle Inventories, Ed. Ecoinvent report No. 8, 508–529.

[160] Quain, S. 2019. The Definitions of "Upstream" and "Downstream" in the Production Process. https://smallbusiness.chron.com/definitions-upstream-downstream-production-process-30971.html. Accessed 16 September 2020.

[161] Rébpublique Tunisienne. 1996. Code du Travail. Loi n° 96-62.

[162] 1999. Règlement interne de la companie des phosphates de Gafsa. In Arrêt n°2644.

[163] Research Center CPG. Généralités sur le phosphate naturel du bassin de Gafsa. General information on rock phosphate in the Gafsa Basin. Internal report, 1.

[164] Research Center CPG. 1999. Rapport: études granulochimique des qualités marchandes du phosphate du bassin de Gafsa. Internal file, 3.

[165] Research Center CPG. 2017. Analyse chimique de l'eau de forage. Internal document. Accessed 7 April 2020.

[166] Research Center CPG. 2017. Rapport chimique des rejets fins laverie 4. Internal document. Accessed 7 April 2020.

[167] Research Center CPG. 2017. Synoptique de la partie brut, lavage et tratement d'eau. Internal document. Accessed 7 April 2020.

[168] Rietbergen-McCracken, J. and Narayan, D. 1996. Participation and social assessment tools and techniques 17796. The International Bank for Reconstruction and Development.

[169] RitchieSpecs Equipment Specification. 2018. Caterpillar D10R Crawler Tractor.

[170] Rockström, J., Steffen, W., Noone, K., Persson, Å., Chapin, F. S., III, Lambin, E., Lenton, T. M., Scheffer, M., Folke, C., Schellnhuber, H. J., Nykvist, B., Wit, C. A. de, Hughes, T., Van der Leeuw, S., Rodhe, H., Sörlin, S., Peter K. Snyder, Costanza, R., Svedin, U., Falkenmark, M., Karlberg, L., Corell, R. W., Fabry, V. J., Hansen, J., Walker, B., Liverman, D., Richardson, K., Crutzen, P., and Foley, J. Planetary Boundaries: Exploring the Safe Operating Space for Humanity.

[171] Rowland, J. H. and Mainiero, R., Eds. 2000. Factors affecting ANFO fumes productions. Proceedings of the 28th Annual Conference on Explosives and Blasting Technique Vol.1. International Society of Explosives Engineers, Cleveland, OH.

[172] Ryan, Y. 2014. Special documentary - Tunisia: The phosphate curse. https://english.alaraby.co.uk/english/features/2014/12/18/special-documentary-tunisia-the-phosphate-curse. Accessed 25 June 2020.

[173] Salhi, B. 2017. Mutations socio-spatiales et environnementales du bassin minier de Gafsa (Sud-ouest de Tunisie): Approche par les outils géomatiques. Dissertation, Université Bretagne Loire.

[174] Sapko, M., Rowland, J. H., Mainiero, R., and Zlochower, I., Eds. 2002. Chemical and Physical Factors that Influence N0x Production During Blasting: Exploratory Study. Proceedings of the 26th Annual Conference on Explosives and Blasting Technique Vol. 2. International Society of Explosives Engineers, Cleveland, OH.

[175] Scholz, R. W., Roy, A. H., Brand, F. S., Hellums, D. T., and Ulrich, A. E., Eds. 2014. Sustainable Phosphorus Management. A Global Transdisciplinary Roadmap. Springer Netherlands, Dordrecht, s.l.

[176] Scholz, R. W., Ulrich, A. E., Eilittä, M., and Roy, A. 2013. Sustainable use of phosphorus: a finite resource. The Science of the total environment 461-462, 799–803.

[177] Schröder, J.J., Cordell, D., Smit, A.L., and Rosemarin, A. Sustainable Use of Phosphorus. EU Tender ENV.B.1/ETU/2009/0025 357. Plant Research International Business Unit Agrosystem; Stockholm Environmental Institute (SEI).

[178] Searates.com. 2020. International container shipping | Online freight marketplace. https://www.searates.com/. Accessed 7 February 2020.

[179] Sellami, A., Sellami Boudawara, T., Hsairi, M., Jlidi, R., and Achour, N. 2007. Incidence des cancers dans le gouvernorat de Sfax 2000-2002. Incidence of cancer in the governorate of Sfax 2000-2002. Hôpital universitaire Habib Bourguiba Sfax, Tunisia.

[180] Shafaie, A. and Heni, W. 2019. Des solutions importantes pour consacrer la transparence et la bonne gouvernance dans le cadre juridique minier tunisien. https://resourcegovernance.org/blog/des-solutions-importantes-pour-consacrer-la-transparence-et-la-bonne-gouvernance-dans-le-cadre. Accessed 12 October 2019.

[181] SKAKO VIBRATION. Caractéristiques Techniques Tambours le lavage type SF_F.

[182] Smida, O. 2012. Etude du comportement des rejets de l'industrie phosphatière du bassin de Gafsa- Metlaoui (M'dhilla, R'deyef, Moularès) dans les conditions de surface. Study of the behaviour of discharges from the phosphate industry in the Gafsa-

Metlaoui basin (M'dhilla, R'deyef, Moularès) under surface conditions., Faculté des sciences de Tunis.

[183] Sociéé tunisienne de l'électricité et de Gaz. 2019. Clients industriels et tertiaires / Tarifs moyenne tension moyenne pression. industrial client: priceso of medium voltage electricty.

[184] Société Tunisienne de transport des produits miniers. 2017. Etat du matériels de la STTPM. Internal document, 11.

[185] Société Tunisienne de transport des produits miniers. 2018. Rapport d'activité de chargement et de transport des produits de phosphates de la CPG selon le secteur pour l'année 2017, 12. Accessed 29 June 2020.

[186] SOTEMU. 2016. Products catalog. https://www.yumpu.com/en/document/read/42899618/products-catalog-sotemu-sa. Accessed 21 April 2020.

[187] SurveyMonkey. Welcome to SurveyMonkey! https://www.surveymonkey.com/dashboard/. Accessed 15 January 2020.

[188] Tartakovsky, D., Stern, E., and Broday, D. M. 2016. Indirect estimation of emission factors for phosphate surface mining using air dispersion modeling. The Science of the total environment 556, 179–188.

[189] The New York Times. 1984. Curfew imposed across Tunisia as riots spread-Archive. The New York Times (1984).

[190] The New York Times. 1987. A Coup is reported in Tunisia. A version of this article appears in print on Nov. 7, 1987, Section 1, Page 3 of the National edition with the headline: A COUP IS REPORTED IN TUNISIA. The New York Times (1987).

[191] Tournis, V. and Rabinovitch, M. 2009. Les ressources naturelles pour la fabrication des engrais : une introduction. Natural ressources for the manufacturing of fertilizer: Introduction. Géologues, Geol162, 37–44.

[192] Transparency International. Corruption Perceptions Index 2018. https://www.transparency.org/cpi2018. Accessed 12 March 2019.

[193] Tuniscope. 2015. Transport du phosphate,la corruption pointe du nez. Transport of phosphate, corruption is here. Tuniscope 2015 (May. 2015).

[194] U.S. Bureau of Mines and U.S. Geological Survey. 1980. Principles of Resrouce/Reserve Classification for Minerals. Geological Survey Circular 831.

[195] U.S. Department of Justice and the Federal Trade Commission. 1997. Horizontal Merger Guidelines (1997). https://www.ecb.europa.eu/pub/pdf/scpwps/ecb.wp2253~cf7b9d7539.en.pdf. Accessed 26 August 2020.

[196] U.S. Department of Justice and the Federal Trade Commission. 2010. Horizontal Merger Guidelines (2010). https://www.justice.gov/atr/horizontal-merger-guidelines-08192010. Accessed 27 August 2020.

[197] U.S. Environmental Protection Agency. Chapter 13: Miscellaneous Sources: Emission unpaved road. In AP 42, Fifth Edition.

[198] U.S. Environmental Protection Agency. 1998. AP-42 11.9 Western Surface Coal Mining. Revision of Emission Factors for AP-42 Section 11.9 Western Surface Coal Mining.

[199] U.S. Geological Survey. 2017. EarthWord–Rock vs. Mineral. https://www.usgs.gov/news/earthword-rock-vs-mineral. Accessed 10 October 2019.

[200] United Nations Development Programme (UNDP). 2018. Human Development Indices and Indicators. United Nations Development Programme (UNDP).

[201] United Nations Environment Programme (UNEP). NaN. Understanding the Long-Term Impacts of Natural Resource Extraction. https://www.unenvironment.org/news-and-stories/story/understanding-long-term-impacts-natural-resource-extraction. Accessed 23 August 2020.

[202] United Nations Environment Programme (UNEP). 2009. Guidelines for social life cycle assessment of products. United Nations Environment Programme, Paris, France.

[203] USGS National Minerals Information Center. 2009. Mineral Commodity Summaries 2008-Phosphate rock. USGS National Minerals Information Center.

[204] USGS National Minerals Information Center. 2010. Mineral Commodity Summaries 2009-Phosphate rock. USGS National Minerals Information Center.

[205] USGS National Minerals Information Center. 2011. Minneral Commodity Summaries 2010-Phosphate rock. USGS National Minerals Information Center.

[206] USGS National Minerals Information Center. 2012. Mineral Commodity Summaries 2011-Phosphate rock. USGS National Minerals Information Center.

[207] USGS National Minerals Information Center. 2013. Mineral Commodity Summaries 2012-Phosphate rock. USGS National Minerals Information Center.

[208] USGS National Minerals Information Center. 2014. Mineral Commodity Summaries 2013-Phosphate rock. USGS National Minerals Information Center.

[209] USGS National Minerals Information Center. 2015. Mineral Commodity Summaries 2014-Phosphate rock. USGS National Minerals Information Center.

[210] USGS National Minerals Information Center. 2016. Mineral Commodity Summaries 2015-Phosphate rock.

[211] USGS National Minerals Information Center. 2017. Mineral Commodity Summaries 2016-Phosphate rock. USGS National Minerals Information Center.

[212] USGS National Minerals Information Center. 2018. Mineral Commodity Summaries 2017-Phosphate rock. USGS National Minerals Information Center.

[213] USGS National Minerals Information Center. 2019. Mineral Commodity Summaries 2018-Phosphate rock. USGS National Minerals Information Center. DOI=10.3133/70202434.

[214] Van Dijk, K. C., Lesschen, J. P., and Oenema, O. 2016. Phosphorus flows and balances of the European Union Member States. The Science of the total environment 542, Pt B, 1078–1093.

[215] Van Kauwenbergh, S. J. September/ 2010. World Phosphate Rock Reserves and Resources.

[216] VBW. 2009. Rohstoffsituation Bayer: Keine Zukunft ohne Rohstoffe. Situation of raw materials in Bavaria: No future without raw materials. Strategien und Handlungsoptionen. Vereinigung der Bayerischen Wirtschaft (VBW).

[217] VBW. 2011. Rohstoffsituation Bayern: Keine Zukunft ohne Rohstoffe. Situation of raw materials in Bavaria: No future without raw materials. Strategien und Handlungsoptionen. Vereinigung der Bayerischen Wirtschaft (VBW).

[218] VBW. 2015. Rohstoffsituation der bayerischen Wirtschaft. Situation of raw materials for the Bavarian economy. Vereinigung der Bayerischen Wirtschaft (VBW).

[219] VBW. 2017. Rohstoffsituation der bayerischen Wirtschaft. Situation of the raw materials for the Bavarian economy. Vereinigung der Bayerischen Wirtschaft (VBW).

[220] Waszkewitz, H. 2018. Tunisia's phosphate mines – between a rock and a hard place. https://globalriskinsights.com/2018/02/tunisia-phosphate-mines-strikes/. Accessed 1 February 2020.

[221] Weatherspark. 2020. Météo habituelle à Metlaoui, Gafsa Tunisie - Weather Spark. Weather at Metlaoui, Gafsa, Tunisia - Weather Spark. https://fr.weatherspark.com/y/58621/M%C3%A9t%C3%A9o-habituelle-%C3%A0-Metlaoui-Tunisie. Accessed 22 June 2020.

[222] Wendling, Z. A., Emerson, J. W., Esty, D. C., Levy, M. A., Sherbinin, A. de, Spiegel, N. R., Pinkerton, V., Boucher, L., Ratté S., Mardell S., Ichihara, M., Battles, J., and Quay, A. N. 2018. Environmental Performance Index. Global metrics for the environment: ranking country performance on high priority environmental issues, New Haven.

[223] World Bank. 2017. World governance indicators. http://info.worldbank.org/governance/wgi/.

[224] World Health Organization. 2018. Noncommunicable Diseases (NCD) Country Profiles, Tunisia.

[225] World Health Organization. 2020. Common pollutants from household heating, cooking and lighting. https://www.who.int/airpollution/household/pollutants/combustion/en/. Accessed 14 April 2020.

[226] World Health Organization. 2020. Ionizing radiation, health effects and protective measures. https://www.who.int/news-room/fact-sheets/detail/ionizing-radiation-health-effects-and-protective-measures. Accessed 14 April 2020.

[227] Yara UK. 2018. Conversion calculator , plant nutrients. https://www.yara.co.uk/crop-nutrition/farmers-toolbox/conversion-calculator/. Accessed 29 June 2020.

[228] Zepf, V., Simmons, J., Reller, A., Ashfield, M., and Rennie, C. 2014. Materials critical to the energy industry - An introduction. BP.

Annex

A. Chapter 1: Global supply of phosphorus and criticality analysis

Table A-1: Literature review of phosphate rock criticality assessment

Title	Focus of the study	Scope	Applied method	Reference
Minerals, critical minerals, and the U.S. economy (2008)	Identification of nonfuel minerals which are "critical" for domestic industry and emerging technologies; assess their global trends in sources and production; examine potential constraints on their availability; identify impacts of restrictions in their supply on the domestic economy	U.S.A	Criticality matrix with two dimensions; - Supply risk: likelihoods of supply restriction - Impact of supply restriction on the national economy	[141]
Rohstoffsituation Bayern: keine Zukunft ohne Rohstoffe (2009) (2011) (Raw material situation in Bavaria: no future without raw material)	Supporting bavarian companies in securing a long-term supply of raw materials. Promoting research into the more efficient use of raw material subtitues, and developing future recycling concepts with the industry.	Bayern	**Quantitative indicators:** - Static lifetime - Global Politic Risk Index - Concentration in 3 countries - Concentration in 3 companies - Price risk a new indicator in 2011 **Qualitative indicators:** - Importance for future technologies - Strategically and political use - Substitutability	[216, 217]

Report on critical raw materials for the EU (2010-2011) (2014)	Establishing the list of critical raw material criticality for the European union: 20 out of 54 raw materials assessed compared to 14 critical materials out of 41 raw materials assessed in 2011.	European Union	Criticality matrix based on two dimensions: - Economic importance: the share of end use of a raw material in Mega-sectors - Supply risk: concentration index in countries with poor governance, multiplied by the recyclability and substitutability	[69–71]
Kritische Rohstoffe für Deutschland: Identifikation aus Sicht deutscher Unternehmen wirtschaftlich bedeutsamer mineralischer Rohstoffe, deren Versorgungslage sich mittel- bis langfristig als kritisch erweisen könnte (2011) Critical Raw Materials for Germany: Identification from the perspective of German companies of economically significant mineral raw materials whose supply situation could prove critical in the medium to long term	Assessing 52 raw materials to identify the most economically important materials to the German industry associated to a high risk of supply within different time horizons. the studied year is 2008	Germany	Criticality matrix divided into 6 sections based on two dimensions: - Vulnerability - Supply Risk for each dimension a set of indicators according to their time relevance were weighted	[67]
Resource Criticality and Commodity Production Projections (2012)	Long-term future trajectory projection for key minerals production in Australia using peak mineral criticality framework to identify areas where technology and policy could contribute to sustainable resource management in Australia.	Australia	Geologic Resource Supply-Demand Model (GeRS-DeMo)	[84]
Methodology of metal criticality determination (2012)	Developing criticality assessment method at three organizational levels: corporate, national and global level considering two different temporal	Corporate, national. and global level	Criticality space based on three dimensions: - supply risk (SR): medium term and long term - environmental implications (EI)	[87]

	perspectives for supply risk: medium term and long term.		- vulnerability to supply restriction	
Analyse kritischer Rohstoffe für die Landesstrategie Baden-Württemberg (2014) Analysis of critical raw materials for the Baden-Württemberg strategy	Defining the most important minerals for the state Badew-Würrtemberg industries and the risk of supply restriction	Federal state: Baden Württemberg. Germany	Criticality Index: based on weighting three criteria: - Economic importance weighting index - Volume importance weighting index - Supply risk weighting index	[124]
Materials critical to the energy industry - An introduction (2014)	Identify 23 materials involved in the energy pathways and their potential constraints on economically sustainable supply or use.	Energy Sector	Classify Supply interruption indicators into 3 impacts categories Low. Medium and High impact. The supply interruption indicators are: - Reserves availability - Trade restriction: political stability, trading agreement, embargo - Ecological impact - Processing: know-how often in the control of a small number of companies. whose influence can constrain supply and certainly affect price - Substitutability - Recyclability	[228]
Study on Data for a Raw Material System Analysis: Roadmap and Test of the fully operational MSA for raw materials final report (2015)	Mapping the flows of material through the EU economy with additional information related to security of supply	European Union	Material system analysis (MSA) defined by three types of parameters: Physical flow. risk of supply and future demand along the entire life cycle of materials	[13]
Criticality Assessment of Metals for Japan's Resource Strategy (2015)	Strategic minerals for Japan were defined according to a background of their growing importance in industry and rising supply risk. 30 minerals were	Japan	Criticality Index: evaluating five risks categories under 12 parameters. The risk categories are: Supply risk. Price risk. Demand risk. Recycling restriction and Potential usage restriction	[96]

	designated "strategic minerals" for Japan.			
Rohstoffe für Zukunfts-technologien (2016) (Raw material for future technologies)	Scenario analysis of the raw materials demand for 42 emerging technologies in the year 2035	Germany	Foresight method to determine the demand for the respective mineral for selected technologies in 2035 related to the global primary production of the mineral in 2013.	[130]
Ressourcenstrategische Betrachtung der Kritikalität von Phosphor (2016) (Resource strategic consideration of the criticality of phosphorus)	Criticality analysis of the main important phosphate product on the global value chain of phosphorus	Globale-Value Chain	Criticality assessment of - Phosphate rock - Phosphoric acid - White phosphorus - Phosphate Derivates - Recycled Phosphorus	[82]
Rohstoffsituation für bayerischen Wirtschaft (2015) (2017) (Situation of raw materials for the Bavarian economy)	Supporting bavarian companies in securing a long-term supply of raw materials. Promoting research into the more efficient use of raw material subtitues, and developing future recycling concepts with the industry.	Bayern	**Quantitative indicators:** - Static lifetime (geological. - Global Politic Risk Index (- Concentration in 3 countries - Concentration in 3 companies **Qualitative indicators:** - Importance for future technologies - Strategically and political use - Substitutability	[218, 219]
Methodology for establishing the EU list of critical raw materials (2017)	Revision of the EC methodology and update the list of critical raw materials compared to the previous versions 27 critical raw materials out of 78 raw materials	European Union	Criticality matrix based on two dimensions: - Economic importance: the share of end use of a raw material in a NACE sectors - Supply risk: country concentration scaled to the country political stability and to trade restriction, multiplied by the recyclability and substitutability	[72]

DERA-Rohstoffliste: Angebotskonzentration bei Metallen und Industriemineralen - Potenzielle Preis- und Lieferrisiken (2012-2016) (2019) DERA raw material list: Supply concentration of metals and industrial minerals - Potential price and supply risks	Geo-concentration of the global market and potential price and risk of supply of metals and mineral for the German industry	Germany	Matrix of Supply risk divided into three group of risk; according two dimensions: - Country concentration (HHR) - Weighted country risks (WGI)	[42–45]

B. Chapter 2: Methodology

- **ReCipe method**

The LCIA phase is the transformation of LCI flows (input-output) into a set of environmental impact categories. Those environmental impact categories are expressed in quantitative indicators [86] (see Figure B-1). The set of impact categories are organized in a specific LCIA method (such as CML. Eco-indicator 99. Cumulative Energy Demand. ReCiPe). For this work. the method to be applied is the "ReCiPe[1] Midpoint (H) [v1.11 2014]" as it combines CML and Eco-indicator 99 in an updated version [1]. Moreover. ReCiPe distinguishes two levels of environmental damage. which are the Midpoint and the Endpoint level.

The midpoint level refers to the conversion of the LCI results (such as fresh-water use. land use. and the emission of CO_2. NO_x. SO_2) into Midpoint impact category indicators (such as water depletion. climate change. and water ecotoxicity). The Endpoint level refers to the conversion of Midpoint impact categories into three Endpoint impact categories, which are; the damage to human health (HH). the damage to ecosystem diversity (ED). and the damage to resource availability (RA) [86]. The conversion is based on an environmental mechanism. The environmental mechanism is a cascade effect that can lead to a certain level of damage to human health or the ecosystem [86].

The ReCiPe method has three characterization scenarios or perspectives defined according to the different sources of uncertainty and the different choices and assumptions. For this work, the hierarchist perspective (H) was used as it is considered to be the default model. and a consensus model often used in scientific modeling [158]. The Hierarchist perspective has a longer time horizon for the environmental mechanism as the individualist perspective. For instance, the time horizon perspective for the environmental mechanism of climate change is 100 years in the Hierarchist perspective and 20 years in the individualist perspective.

According to ReCiPe. 2008 report: "Perspectives are used to group similar types of assumptions and choices. For instance:

[1] The acronym represents the initials of the main contributors and collaborators institutes for the development and design of the method: **R**IVM and Radboud University, **C**ML, and **P**Ré consulting[86].

- ***Individualist perspective (I)*** *is based on the short-term interest. undisputed impact types. technological optimism as regards human adaptation.*
- ***Hierarchist perspective (H)*** *is based on the most common policy principles with regards to time-frame and other issues.*
- ***Egalitarian perspective (E)*** *is the most precautionary perspective. taking into account the longest time-frame. impact types that are not yet fully established. but for which some indication is available." [86]*

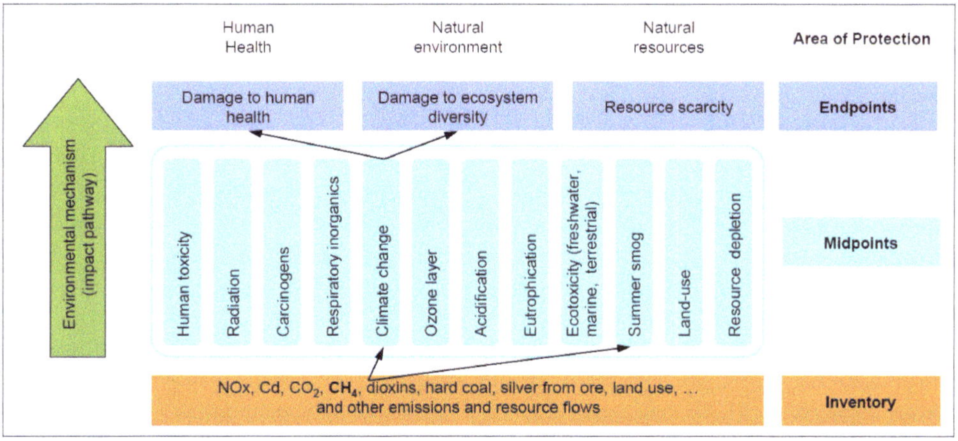

Figure B-1: Relationship between inventory results. midpoint indicators. and endpoint indicators in the LCIA [74]

Table B-1: Impact categories and characterization factor in ReCiPe [86]

Impact category	Abbr.	Unit	Characterization factor name	Abbr.
Climate change	CC	kg (CO_2 to air)	global warming potential	GWP
Ozone depletion	OD	kg (CFC-115 to air)	ozone depletion potential	ODP
Terrestrial acidification	TA	kg (SO_2 to air)	terrestrial acidification potential	TAP
Freshwater eutrophication	FE	kg (P to freshwater)	freshwater eutrophication potential	FEP
Marine eutrophication	ME	kg (N to freshwater)	marine eutrophication potential	MEP
Human toxicity	HT	kg (14DCB to urban air)	human toxicity potential	HTP
Photochemical oxidant formation	POF	kg (NMVOC6 to air)	photochemical oxidant formation potential	POFP
Particulate matter formation	PMF	kg (PM10 to air)	particulate matter formation potential	PMFP
Terrestrial ecotoxicity	TET	kg (14DCB to industrial soil)	terrestrial ecotoxicity potential	TETP
Freshwater ecotoxicity	FET	kg (14DCB to freshwater)	freshwater ecotoxicity potential	FETP
Marine ecotoxicity	MET	kg (14-DCB7 to marine water)	marine ecotoxicity potential	METP
Ionizing radiation	IR	kg (U235 to air)	ionizing radiation potential	IRP
Agricultural land occupation	ALO	m^2xyr (agricultural land)	agricultural land occupation potential	ALOP
Urban land occupation	ULO	m^2xyr (urban land)	urban land occupation potential	ULOP
Natural land transformation	NLT	m^2 (natural land)	natural land transformation potential	NLTP
Water depletion	WD	m^3 (water)	water depletion potential	WDP
Mineral depletion	MRD	kg (Fe)	mineral depletion potential	MDP
Fossil energy depletion	FD	kg (oil)	fossil depletion potential	FDP

- **Social Impacts Weighting method**

For calculating social impacts for a specific product system, since social impacts assessment is still investigated. and no generally accepted method has been developed so far. the so-called "Social Impacts Weighting Method" is rather rudimentary with typically impact categories corresponding to one indicator [64].

The reference unit of each impact category is medium risk hours. Therefore, the risk level of each impact category is scaled to medium risk applying an impact factor as indicated in Table B-3.

Table B-2: Risk levels and their corresponding Impact factors [64]

Risk level	Factor
Very low risk	0.01
Low risk	0.1
Medium risk	1
High risk	10
Very high risk	100
No risk	0.1
No data	0.1

- **Stakeholder analysis**

Table B-3: Stakeholder classification and the management action

Stakeholder category	Interst/power level	Management action
Category 1	Low/Low	Little effort
Category 2	Low/High	Keep satisfied
Category 3	High/Low	Keep informed and maintain interested
Category 4	High/High	Manage very closely

Table B-4: Stakeholder classification

Objective:	**Define the degree of interest and power of the CPG stakeholders**				
Interest:	Assess the level of interest that this stakeholder attaches to the success of the business				
Scale of interest:	**0: very low**	**1: low**	**2: intermediate**	**3: high**	**4: very high**
Power:	Assess the degree of positive or negative influence that this stakeholder can exert on the company				
Scale of power:	**0: very low**	**1: low**	**2: intermediate**	**3: high**	**4: very high**
Source used:	https://www.mybeeye.com/outils-gestion/parties-prenantes				
	Norm: ISO 9001 :2015				

Table B-5: Stakeholder classification according to the participant at the CPG

Stakeholder	**Interest**	**Power**
Shareholders		
State	4	4
Employees		
White collar	4	4
Blue collar	4	2
Suppliers		
Electricity	1	2
Fresh water SONED	2	1
Machine	2	1
Chemicals	2	1
Service providers		
Rail transportation: SNCFT	3	3
Road transportation : STTPM	4	4
Local population	3	2
Tunisian society	1	1
Local authorities		
Gouvernorate	3	3
Municipality	3	1
Syndicate	4	4
Civil society		
Association human rights	1	1
Association for the environment	3	3
Association for culture and sports	3	1
Client :		
Groupe Chimique GCT	4	4
TIFERT	4	4
Medias (local and national)x	2	2

National and regional research institution		
Universities	2	1
Research insitution	2	2
Local organization and agencies		
Hospitals	3	1
Regional tourism office	1	1
Regional office for agriculture development CRDA	1	1
National employment agency :(ANETI)	1	1
Banks	3	3
Local transport company : Société de transport régional de Gafsa (STRG)	3	2
Municipal Wastewater treatment company ONAS	1	1

Figure B-2: Stakeholders power interest grid

Gender	♂ 49,9%	♀ 50,1%

Age Group	15-19 years	20-29 years	30-39 years	40-49 years	50-59 years	>=60
Male	9%	23%	22%	17%	15%	14%
Female	10%	24%	22%	17%	14%	13%

Education	School 6-14 years	University t 19-24 years	Anaphalbetism: 10 years and over	Internet use: 15 years and over
	97%	43%	20%	43%

- 95% of confidence level and 10% error margin : Target Size 230 people
- Response rate estimated to 60%
- Population sample Structure:
  50% F and 50% M.
  20% 20-29 years/ **20%** 30-39/ **15%** 40-49/ **10%** 50-60

- Size of total Population 209 076
- Size of literate Population: 122 676

Figure B-3: The characterization of local population in Gafsa and the population sample structure

Table B-6: Social impacts, their indicator and the corresponding question for the survey

Category	Indicators	Questions
Health	Quality of health care	Can you rate the reason behind leaving the region? Unemployment. Health/ access to clean water/ Pollution/Security. Other reason :....
	Diseases related to the phosphate activities	How high is the impact of Phosphate mining on the following public health in the region? Air pollution related diseases. Water pollution related diseases. Cancer
Water	Drinking water access	Please rate the quality of drinking water coverage How frequent is the interruption of water in households per month. week and day?
	Industrial water Use	How frequent is the interruption of water in households per month. week and day? According to your personal opinion please estimate the risk of the following impact of Phosphate mining on the water resources Water depletion/ water pollution
Economic opportunity	Unemployment rate in the region	Please rate the reason behind leaving the region: Unemployment. Health. Pollution. Security
	Local economic opportunity	Please assess the quality of influence of phosphate mining on the following economic aspects: General Economic development of the region/ Attractiveness to investment/ Agriculture/ Tourism
Environmental load And quality of life	Environmental pollution	Please estimate the risk of the following impact of Phosphate mining on surrounding villages: During mining operation. the company removes native vegetation The soil pollution by Phosphate mining destroys arable lands
	Livability of the area	Please estimate the risk of the following impact of Phosphate mining on surrounding villages: The use of explosive damages the houses/ The disposal of mining waste is harming the landscape of the area

Table B-7: Ordinal scale questions (1-9) survey among the local population

- 1- Please assess the quality of influence

	Very bad	Bad	Neutral (neither bad nor good)	Good	Very good	I don't know
The influence of phosphate mining on Gafsa 's economic development is						
The influence of the Phosphate mining on the attractiveness to investment in the region is						
The influence of Phosphate mining on agriculture is						
The influence of Phosphate mining on tourism is						

- 2- Can you rate the reason behind leaving the region

	Very Relevant	Relevant	More or Less relevant	Not very relevant	Not relevant at all	I don't know
Absence of employment opportunities						
Environmental pollution						
Unsafety: crime. terrorism						

Difficult access to clean water supply and sanitation						
Difficult access to health care services						

- 3- In your personal opinion. how high is the impact of phosphate mining on the following public health problem in the region

	No Impact		**Very low Impact**	**Low Impact**	**High Impact**	**Very high Impact**	**I don't know**
Air pollution related health problems							
Water pollution related health problem							
Cancer							

- 4- Please rate the quality of drinking water coverage:

	Very Good	**Good**	**Average**	**Bad**	**Very bad**	**I don't know**
The quality of drinking water						

- 5- How frequent is the interruption of water in households per month, week and day?

	Annually	Per Month	Weekly	Daily	
No interruption	**1- 3 times per year**	**Low frequency: 1-3 times per month**	**High frequency: At least 2 times per week (twice or more a week)**	**Very High frequency: At least once per day (once or more a day)**	**I don't know**

- 6-According to your personal opinion please estimate the risk of the following impact of phosphate mining on the water resources

	No Risk	Very Low Risk	Low Risk	High Risk	Very high Risk	I don't know
Phosphate mining causes the scarcity of water resource in the region						
Phosphate mining Pollute water bodies by industrial effluent						

- 7- According to your personal opinion. please estimate the risk of the following impact of Phosphate mining on surrounding villages

	No Risk	Very Low Risk	Low Risk	High Risk	Very high Risk	I don't know
During the preparation of the mining site the road construction obliges nearby community to leave their land						
During mining operation the company removes native vegetated area						
During the blasting of the overburden layer the explosion creates soil movement beneath the houses						
The disposal of mining waste is harming the landscape of the area						
The soil pollution by Phosphate mining destroys arable lands						

- 8- As a part of the local population, how informed do you feel about the following topics?

	Well informed	Somehow informed	Not enough informed	Not informed	I don't know
The Project for new mining site construction					
Phosphate production process					
Risks associated to mining waste disposal					
Development projects in the region funded by the CPG Company					
Financial results					

- 9- As a part of the local population, how well involved in the company's decision?

well involved	Somehow involved	Not enough involved	Not involved	I don't know

- 10-Gender

  - female
  - male
  - other/unknown

C. Chapter 3: Revision of the global phosphate reserves

Table C-1: **Updates of the measured and demonstrated reserves in Gafsa mining region according to CPG geological department 2018 [136]. Quantities expressed in million metric tons**

	01.01.2011	**01.01.2017**	**01.01.2018**		
Reserves category	Measured	Measured	Indicated	Measured	Demonstrated
Sector1	168.75	155.919	153.158	137	290.158
Sector2	155.25	150.148	149.066	40	189.066
Sector3	59.15	53.055	52.062	104	156.062
Sector4	126.5	126.461	126.259	40	166.259
Sector5	50.8	49.483	48.723	44	92.723
Sector6	76.65	72.905	72.040	36	108.016
Total sectors	**637.1**	**607.971**	**601.314**	**401**	**1002.314**
Project1	107.1	176	176	264	440
Project2	20	20	20	96	116
Project3	2,500	2,500	2,500	2,500	5,000
Total Projects	**2,627.1**	**2,696**	**2,696**	**2,860**	**5,556**
Total Reserves	**3,264.2**	**3,303.971**	**3,297.314**	**3,261**	**6,558.314**

Table C-2: **Evolution of the global reserves of phosphate rock, global production of phosphate rock and the Resources/Consumption ration based on the data of the USGS and data [203–213] of the CPG for the year 2017-2018 [136]**

	2011	**2012**	**2013**	**2014**	**2015**	**2016**	**2017**	**2018**
Reserve PR (MMT)	71,152	67,238	66,758	64,117	68,705	67,825	73,108	79,174
Production PR (MMT)	198.080	213.102	225.160	218.190	241.120	255.160	269.495	279.550
Ratio (years)	359.208	315.520	296.491	293.858	284.941	265.813	259.391	271.914
Growth rate of reserves	9%	-6%	-1%	-4%	7%	-1%	8%	8%
Growth rate of consumption	10%	8%	6%	-3%	11%	6%	6%	4%
AAG (Reserve)	3%							
AAG (Consumption)	6%							
Slope (ratio)	-0.07							

Table C-3: European Importation: quantity and countries source of exportation based on the data from ITC 2016

EU Countries	Total imports (tons)	Imports outside Europe (tons)	Exporting countries
Poland	1 313 099	1 312 268	Morocco. Senegal. Algeria
Lithuania	1 281 393	1 281 391	Russia. Morocco
Belgium	864 363	779 122	Russia. Morocco
Bulgaria	481 143	480 536	Morocco. Senegal. Jorand. Egypt
Netherlands	279 651	277 021	Isragel. Morocco
Greece	177 248	177 248	Morocco. Algeria
France	179 107	171 825	Morocco. Algeria
Germany	110 303	108 147	Israel
Romania	99 985	98 746	Egypt
Portugal	59 408	58 979	Morocco. Israel
Austria	48 442	47 448	Egypt. Jordan
Croatia	47 918	46 140	Algeria
Spain	42 148	34 143	Morocco. Israel. Senegal. Egypt. Algeria. Tunisia
Estonia	458	283	Russia
United Kingdom	801	221	USA. China
Italy	130 669	104	Egypt. Algeria. Morocco
Denmark	1 692	25	China
Sweden	5 081	21	New Zealand. USA
Cyprus	3	0	
Czechia	350	0	
Finland	0	0	
Hungary	515	0	
Ireland	368	0	
Latvia	274	0	
Luxembourg	105	0	
Malta	0	0	
Slovakia	155	0	
Slovenia	4 416	0	
Total imported tons	**5 129 095**	**4 873 668**	

D. Chapter 6: Environmental life cycle analysis of phosphate rock mining

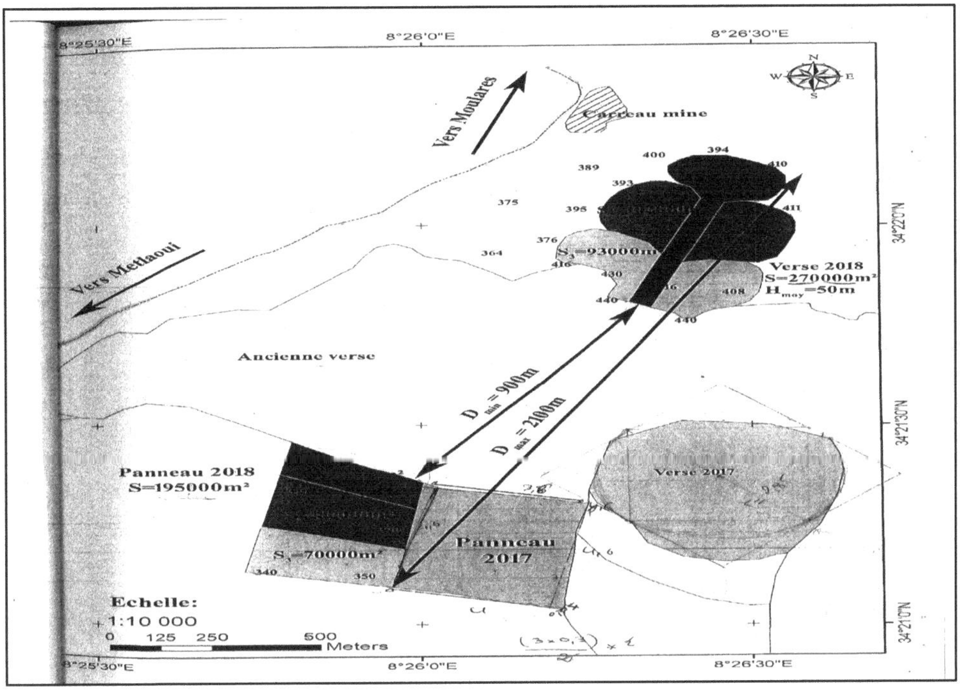

Figure D-1: Planning map for the mining operation at Metlaoui, for the year 2018 [62]

Table D-1: P content in crude phosphate rock ore of Morocco, USA, and Tunisia

	Content	Source
P in P_2O_5	43.645%	[227]
P_2O_5 in Apatite	42%	[93]
P content in Apatite	18%	43.64*0.42
P content in crude ore in Morocco (27% P_2O_5)	12%	[159]
P Content in crude ore in USA (9% P_2O_5)	4%	[159]
P Content in crude ore in Tunisia (24% P_2O_5)	11%	[164]

Figure D-2: Example of satellite picture of the beneficiation plant at Moulares(a) [33] and Kef Eddour (b) [32]. Research Center (c) [30]. Maintenance and reparation center (d) [31]

Table D-2: List of equipment operative at the mining site of Metlaoui in 2017

Function	Name	Brand	Type	Engine Model	Power (KW)	Operating weight: (metric ton)	Load capacity (metric ton or m^3)	Fuel use (L/h)	Source
Drilling	Rotary Blasthole Drills	Atlas COPCO	DM 45	CAT C15		35- 41 t	—		[6]
Excavation	Bulldozer	CAT	D.10 T	Cat® C27 ACERT™	433	66.451 t	18.5 m^3	81.4-100.2	[25]
	Bulldozer	CAT	D.10 R	Cat® 3412E	425.1	65.4035 t	22 m^3	76-93	[169]
Loading	Hydraulic backhoe excavators	Komatsu	PC. 3000	Komatsu SSA12V159	940	250-258 t	15 m^3	172	[122]
	Hydraulic backhoe excavators	Komatsu	PC. 5500	Komatsu (2x) SSA12V159	2 x 940	533 - 552 t	29 m^3	328	[123]
Transport	Dumper 170 T	CAT	789 C	Cat® 3516B EUI	1320	317. 515	177 t	105.9-141.2	[26]
	Dumper 170 T	CAT	789 D	Cat® 3516C EUI	1468	324 319	181 t	112.4-149.9	[27]

Table D-3: Transportation of different phosphate product and byproduct for the year 2017 [185]

Transport parameters	Unit	Crude phosphate rock	Barren stock	Beneficiated phosphate rock	Coarse reject of the beneficiation plant
Distance between different sites	km	446.8	22.2	55.4	64
Throughput	t per year	16451411	1897802	3841298	2676419
Payload distance	tkm/year	176630505	4146691.4	6012279	11750492
Truckload capacity	T	85	180	85	85
Total distance	truck-km/year	2078006	23037.2	70733	138241
Distance/truck	km/truck*year	140183	3291	544	1063

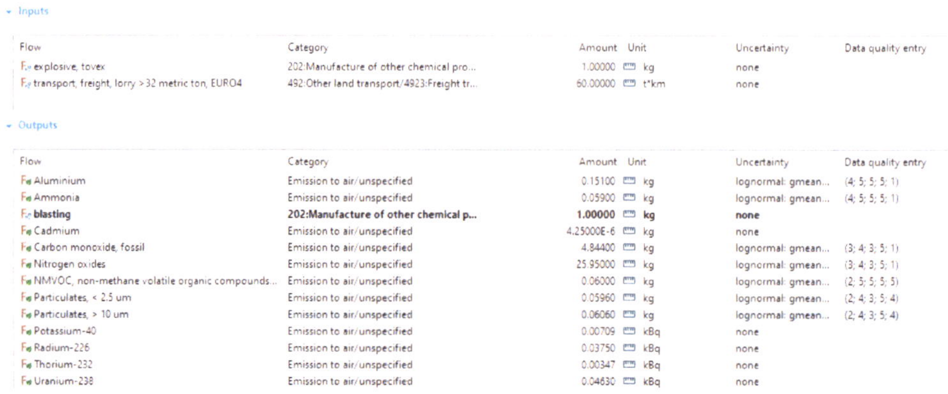

Inputs

Flow	Category	Amount	Unit	Uncertainty	Data quality entry
explosive, tovex	202:Manufacture of other chemical pro...	1.00000	kg	none	
transport, freight, lorry >32 metric ton, EURO4	492:Other land transport/4923:Freight tr...	60.00000	t*km	none	

Outputs

Flow	Category	Amount	Unit	Uncertainty	Data quality entry
Aluminium	Emission to air/unspecified	0.15100	kg	lognormal: gmean...	(4; 5; 5; 5; 1)
Ammonia	Emission to air/unspecified	0.05900	kg	lognormal: gmean...	(4; 5; 5; 5; 1)
blasting	**202:Manufacture of other chemical p...**	**1.00000**	**kg**	**none**	
Cadmium	Emission to air/unspecified	4.25000E-6	kg	none	
Carbon monoxide, fossil	Emission to air/unspecified	4.84400	kg	lognormal: gmean...	(3; 4; 3; 5; 1)
Nitrogen oxides	Emission to air/unspecified	25.95000	kg	lognormal: gmean...	(3; 4; 3; 5; 1)
NMVOC, non-methane volatile organic compounds...	Emission to air/unspecified	0.06000	kg	lognormal: gmean...	(2; 5; 5; 5; 5)
Particulates, < 2.5 um	Emission to air/unspecified	0.05960	kg	lognormal: gmean...	(2; 4; 3; 5; 4)
Particulates, > 10 um	Emission to air/unspecified	0.06060	kg	lognormal: gmean...	(2; 4; 3; 5; 4)
Potassium-40	Emission to air/unspecified	0.00709	kBq	none	
Radium-226	Emission to air/unspecified	0.03750	kBq	none	
Thorium-232	Emission to air/unspecified	0.00347	kBq	none	
Uranium-238	Emission to air/unspecified	0.04630	kBq	none	

Figure D-3: Input-Output table of the process "Blasting phosphate-TN" in openLCA

Table D-4: Particulate matter emission factor g/metric ton of crude phosphate rock according to the EEA calculation model [75, 76], the composition of TSP (%), and Gafsa mining weather conditions

Process step	Emission Factor g/metric ton			Data source
	TSP%	**PM-10%**	**PM-2.5%**	
Blasting/drilling	100%	53%	52%	CPG Internal data[60, 62] [53, 55] US EPA,1998[198]
	15	**8.1**	**8.0**	
Material processing: Dry crushing and grinding without filtering	100%	35%	5%	CPG Internal data, US EPA,1998[198]
	2.1	**0.77**	**0.09**	
Transportation between sites: transport of crude PR and transport of overburden	100%	29%	3%	EPA unpaved road[197]
	6,575	**1,889**	**191**	
Material handling: trucks and Dumper loading and unloading	100%	47%	7%	CPG Internal data[53, 60] /USEPA.,1998[198]
	0.71	**0.33**	**0.051**	
Total Emission factor excluding the blasting/drilling	100	29%	3%	
	6,578	**1,890**	**191**	

Figure D-4: Picture of water and soil contamination by slurry: dark water at the river Wad Thelja between Redeyef and Metlaoui (a) and (b) Copyright Giovani Camici, Google maps/ slurry leakage around M'dhila beneficiation plants (c) and (d)

Table D-5: Chemical concentration in the wastewater fraction of the slurry and their related emissions due to the production of 1kg beneficiated phosphate rock based on the data from Smida (2012) [182]

Chemical	Unit	Concentration per kg wastewater	Emission per 1kg benef. PR
Fluorine (F)	mg/kg	47	89.47
Sulfate (SO_4^{2-})	mg/kg	2570	4,892
Phosphorus (P)	mg/kg	24.8	47.20
Nitrate (NO3)	mg/kg	17.7	33.70
Nitrite (NO2)	mg/kg	15.3	29.12
Cadmium (Cd)	mg/kg	0.01	0.019
Lead (Pb)	mg/kg	0.001	0.0019
Zinc (Zn)	mg/kg	0.053	0.10

Table D-6: Chemical concentration in the slime fraction of the slurry and their related emissions due to the production of 1kg beneficiated phosphate rock based on the data from Smida (2012) [182]

	Unit	Concentration per kg slime	Emission per 1kg benef PR
Cd	**mg/kg**	25.89	7.12
Pb	**mg/kg**	4.37	1.20
U	**mg/kg**	30.16	8.30
Zn	**mg/kg**	271.6	74.75

Figure D-5: Photo of a drying unit at CPG (Own photo during data collection visit to the drying plant)

Table D-7: Inputs/outputs of the production of 1 kg P_2O_5

Input	**Unit**	**Total**	**Mining**	**Benefici-ation**	**Drying**
Infrastructure					
Occupation mineral extraction site (industrial area)	m^2*a	1.02E-03	1.02E-03		
Transformation, to mineral extraction site (from desert)	m^2	1.70E-01	1.70E-01		
Occupation, construction building DMM, Research Center Administration (5 years)	m^2	4.61E-05	4.61E-05		

Occupation for office and maintenance lifetime 20 years	m^2	1.15E-05	1.15E-05		
Buildings crushing, steel	m^2	1.70E-05	1.70E-05		
Building beneficiation plant	m^2	9.43E-06		9.43E-06	
Occupation, construction site (5 years construction)	m^2*a	2.12E-06		2.12E-06	
Occupation, industrial site (20 years occupation time)	m^2*a	5.30E-07		5.30E-07	
Water recycling Plant	item	9.97E-10		9.97E-10	
Conveyor belt (9km, 25 life-time; 1000t/h, 3500h/year)	m	5.37E-06	4.29E-07	4.94E-06	
Machine					
Blasthole Drills DM45 Atlas Copco (35-41 tons, 20yr life-time)	kg	5.89E-06	5.89E-06		
Hydraulic excavator (500 tons, 20yr lifetime)	Item	4.42E-10	4.42E-10		
Bulldozer (66 tons, 20yr lifetime)	kg	3.89E-05	3.89E-05		
Dumper (170 tons, 20yr lifetime)	kg	1.50E-04	1.50E-04		
Truck (85 tons, 15yr lifetime)	Item	4.42E-10	4.42E-10		
Filtering band	m	1.13E-05		1.13E-05	
Equipment for washing plant such as scrubber, hydrocy-lone, screening	Kg	1.15E-04		1.15E-04	
Equipment for drying	kg	7.95E-04			7.95E-04
Maintenance for Machinery (19 Machines, 20yr lifetime)	Item	2.80E-09	2.80E-09		
Movement of soil					
Phosphorus, as P (18% in Ap-atite, 11% in crude ore, in ground)	kg P	4.58E-01	4.58E-01		
Fluorine,4.5% in Apatite, 2,4% in crude ore, in ground	kg F	1.00E-01	1.00E-01		
Operating Resources					
Explosive ANFO	kg	9.04E-03	9.04E-03		
Explosive N30	kg	1.65E-04	1.65E-04		
Total explosive	kg	9.20E-03	9.20E-03		
Diesel (used in machinery, 45.6 MJ per kg diesel)	MJ	7.37E-01	7.25E-01	1.26E-02	

Electricity (Industrial , Medium voltage)	KWh	4.00E-02	6.89E-04	3.93E-02	
Heavy fuel oil n°2, 41 MJ per kg	MJ	2.33E+00			2.33E+00
Total energy wet benef.	MJ	8.81E-01	7.27E-01	1.54E-01	
Lubricant oil	kg	7.79E-04	7.17E-04	6.22E-05	
Industrial Water	m^3	1.17E-02		1.17E-02	
Flocculant: Semifloc	kg	2.52E-04		2.52E-04	
Transport					
Transport of Barren rock (with dumper)	kg-km	3.53E+00	3.53E+00		
Transport of crude phosphate (with truck)	kg-km	1.50E+02	1.50E+02		
Transport equipment (500 t), road (500 km)	kg-km	2.13E-01	2.13E-01		
Transport equipment, (500 t) freight ship (5000 km)	kg-km	2.13E+00	2.13E+00		
Transport Explosives from the SOTEMU, road (60 km)[3]	kg-km	5.52E-01	5.52E-01		
Transport of beneficiated rock lorry	kg-km	5.99E+00		5.99E+00	
Transport of waste	kg-km	1.17E+01		1.17E+01	
Output	**Unit**	**Total**	**Mining**	**Benefici-ation**	**Drying**
Emission to air					
Carbon Monoxide	kg	4.46E-02	4.46E-02		
Nitrogen Monoxide	kg	2.39E-01	2.39E-01		
Fluorine	kg	1.28E-05			1.28E-05
Water vapor	kg	5.41E-01			5.41E-01
Particulate Matter					
Particulates, <10 um	kg PM <10	2.16E-03	2.12E-03		3.92E-05
Particulates, > 2.5 um, and < 10um	kg PM>2.5<10	1.13E-02	1.12E-02		1.05E-04
Particulates, < 2.5 um	kg PM<2.5	2.79E-03	1.63E-03		1.16E-03
Radioactive					
Uranium-238	kBq	1.49E-02	1.49E-02		
Thorium-232	kBq	1.12E-03	1.12E-03		
Radium-226	kBq	1.21E-02	1.21E-02		
K-40	kBq	2.28E-03	2.28E-03		

Emission to water					
Fluorine	mg/kg	3.09E+02		3.09E+02	
Sulfate	mg/kg	1.69E+04		1.69E+04	
Phosphate PO4	mg/kg	1.63E+02		1.63E+02	
Nitrate NO3	mg/kg	1.16E+02		1.16E+02	
Nitrite NO2	mg/kg	1.00E+02		1.00E+02	
Cd	mg/kg	6.56E-02		6.56E-02	
Pb	mg/kg	6.56E-03		6.56E-03	
Zn	mg/kg	3.48E-01		3.48E-01	
Emission to soil					
Cd	mg/kg	2.46E+01		2.46E+01	
lead	mg/kg	4.15E+00		4.15E+00	
U	mg/kg	2.86E+01		2.86E+01	
Zn	mg/kg	2.58E+02		2.58E+02	
U 234	Kbq/kg	4.29E+01		4.29E+01	
U238	Kbq/kg	1.43E+01		1.43E+01	

Table D-8: Total Particulate Matter (TPM) calculation

	Emission Factor (kg/kg)	**Emission of HFO (kg/MJ)**	**Consumption of heavy fuel oil per kg (MJ/kg)**	**Emission of the consumed HFO (kg/MJ/kg)**	**Emission of dryer (kg/kg)**	**Total production of dry benef. PR (kg/year)**	**Total emission of PM (kg/year)**
N° column	A	B	C	D	E	F	J
Source of data	EPA 1993	Ecoinvent Report n°8 Phosphate rock.2007	CPG data. 2016-2017	B*C	A*D	CPG data2016-2017	E*F
2016	0.0004	1.73E-05	0.66	1.15E-05	3.88 E-04	2.3 E+08	8.96 E+04
2017	0.0004	1.73E-05	0.73	1.26E-05	3.87 E-04	1.5 E+08	5.83 E+04
Average	**0.0004**	**1.73E-05**	**0.69**	**1.20E-05**	**3.88 E-04**	**1.9 E+8**	**7.39 E+04**

Table D-9: Water depletion impact of the product system "Wet beneficiation phosphate rock, as P_2O_5-TN". Expressed in m^3/kg P_2O_5

Product system	Total amount	Direct contribution	Main unit process	Amount	Main background process	Amount
Wet beneficiation phosphate rock as P_2O_5-TN	**0.963**	0.012	Market of Wastewater treatment facility	0.537		
			Mining operations phosphate rock as P_2O_5-TN	0.25	Blasting-TN	0.19
			Market of conveyor belt	0.13		

Table D-10: Freshwater eutrophication impact category of the product system "Wet beneficiation phosphate rock, as P_2O_5-TN". Expressed in P eq/kg P_2O_5

Product system	Total amount	Direct contribution	Main unit process	Amount	Main background process	Amount
Wet beneficiation phosphate rock as P_2O_5-TN	**1.50E-04**	5.37E-05	Market of wastewater treatment facility	5.96E-05		
			Mining operations phosphate rock as P_2O_5-TN	1.65E-05	Blasting-TN	1.39E-05
			Market of conveyor belt	1.60E-05		

Table D-11: Freshwater ecotoxicity impact category of the product system "Wet beneficiation phosphate rock, as P_2O_5-TN". Expressed in 1,4 DB eq/kg P_2O_5

Product system	Total amount	Direct contribution	Main unit process	Amount	Main background process	Amount
Wet beneficiation phosphate rock as P_2O_5-TN	**5.22E-03**	9.21E-05	Market of wastewater treatment facility	3.22E-03		
			Market of conveyor belt	7.70E-04		
			Mining operations phosphate rock as P_2O_5-TN	9.40E-04	Blasting-TN	6.70E-04

Table D-12: Terrestrial acidification impact category of the product system "Wet beneficiation phosphate rock, as P_2O_5-TN". Expressed in kg SO_2/kg P_2O_5

Product system	Total amount	Main unit process	Amount	Main background process	Amount
Wet beneficiation phosphate rock, as P_2O_5-TN	**1.37E-01**	Mining operations phosphate rock as P_2O_5-TN	1.36E-01	Blasting-TN	1.35E-01

Table D-13: Terrestrial ecotoxicity impact category of the product system "Wet beneficiation phosphate rock, as P_2O_5-TN". Expressed in kg 1,4 DB eq/kg P_2O_5

Product system	Total amount	Direct contribution	Main unit process	Amount	Main background process	Amount
Wet beneficiation phosphate rock as P_2O_5-TN	1.84E-03	1.78E-03	Mining operations phosphate rock as P_2O_5-TN	6.28E-05	Blasting-TN	3.51E-05

Table D-14: Arable Land use impact categories of the product system "Wet beneficiation phosphate rock, as P_2O_5-TN".

Product system	Impact category	Total amount	Main Unit process	Amount	Main background process	Amount
Wet beneficiation phosphate rock as P_2O_5-TN	**Agricultural land occupation (m^2/yr)**	6.95E-03	Mining operations phosphate rock as P_2O_5-TN	3.36E-03	Blasting-TN	2.75E-03
	Natural land transformation (m^2)	7.62E-05		5.57E-05		2.54E-05

Table D-15: Climate change impact category of the product system "Wet beneficiation phosphate rock, as P_2O_5-TN". Expressed in kg CO_2 eq./kg P_2O_5

Product system	Total amount	Main unit process	Amount	Main background process	Amount
Wet beneficiation phosphate rock as P_2O_5-TN	**3.74E-01**	Mining operations phosphate rock as P_2O_5-TN	1.81E-01	Diesel burned in agriculture machine-TN	6.00E-2
				Blasting-TN	9.77E-02
				Transport on site-TN	1.42E-02
		Market wastewater treatment facility	1.40E-01		
		Market for Electricity medium voltage-TN	2.28E-02		
		Market conveyor belt-GLO	2.01E-02		
		Transport beneficiated phosphate rock and Coarse	1.63E-03		

Table D-16: Ionizing radiation impact category of the product system "Wet beneficiation phosphate rock, as P_2O_5-TN". Expressed in kg 235U eq/kg P_2O_5

Product system	Total amount	Main unit process	Amount	Main background process	Amount
Wet beneficiation phosphate rock as P_2O_5-TN	**2.5E-02**	Mining operations phosphate rock as P_2O_5-TN	1.75E-02	Diesel burned in Agriculture machine	4.00E-03
				Blasting-TN	5.67E-03
				Transport on site-TN	1.07E-03
		Market wastewater treatment facility-GLO	5.80E-03		
		Market for conveyor belt-GLO	1.00E-03		

Table D-17: Particulate matter formation impact category of the product system "Wet beneficiation phosphate rock, as P_2O_5-TN". Expressed in kg PM10 eq /kg P_2O_5

Product system	Total amount	Main unit process	Amount	Main background process	Amount
Wet beneficiation phosphate rock as P_2O_5-TN	**7.18E-02**	Mining operations phosphate rock as P_2O_5-TN	7.13E-02	Blasting-TN	5.29E-02
				Diesel burned in Agriculture machine	2.40E-04
				Transport on site-TN	4.08E-05

Table D-18: Photochemical oxidant formation impact category of the product system "Wet beneficiation phosphate rock, as P_2O_5-TN". Expressed in kg NMVOC eq /kg P_2O_5

Product system	Total amount	Main unit process	Amount	Main background process	Amount
Wet beneficiation phosphate rock as P_2O_5-TN	**2.43E-01**	Mining operations phosphate rock as P_2O_5-TN	2.42E-01	Blasting	2.41E-01

Table D-19: Human toxicity impact category of the product system "Wet beneficiation phosphate rock, as P_2O_5-TN". Expressed in Kg 1,4 DB eq /kg P_2O_5

Product system	Total amount	Direct contribution	Main unit process	Amount	Main background process	Amount
Wet beneficiation phosphate rock as P_2O_5-TN	**2.47E-01**	**1.50E-02**	Mining operations phosphate rock as P_2O_5-TN	1.08 E-01	Blasting-TN	1.60 E-02
					Diesel burned in agriculture machine-GLO	5.16 E-03
					Transport on site-TN	4.67 E-03
					Mining operations direct contribution	6.20 E-02
			Market wastewater treatment facility-GLO	9.50 E-02		
			Market for conveyor belt-GLO	2.48 E-02		

Table D-20: Environmental impact contribution of each life cycle stage

Environmental Impact category	Unit	Mining operations	Wet beneficiation	Drying	Total benef. PR Wet	Total benef. PR Dry
Climate change	kg $CO_{2\ eq}$	1.81E-01	1.94E-01	2.17E-01	**3.75E-01**	**5.92E-01**
Human toxicity	kg 1,4DB $_{eq}$	1.08E-01	1.39E-01	5.38E-02	**2.47E-01**	**3.01E-01**
PM10	kg $PM10_{eq}$	7.13E-02	5.20E-04	8.32E-03	**7.18E-02**	**8.01E-02**
Ozone formation	kg NMVOCs	2.43E-01	7.50E-04	4.38E-02	**2.43E-01**	**2.87E-01**
Ionizing radiation	kg ^{235}U	1.75E-02	7.41E-03	1.31E-02	**2.49E-02**	**3.80E-02**
Water depletion	m^3	2.51E-01	7.12E-01	1.66E-01	**9.63E-01**	**1.13E+00**
Freshwater ecotoxicity	kg 1,4DB $_{eq}$	9.40E-04	4.28E-03	9.70E-04	**5.22E-03**	**6.19E-03**
Freshwater eutrophication	kg P $_{eq}$	1.97E-05	1.30E-04	3.00E-05	**1.50E-04**	**1.80E-04**
Terrestrial acidification	kg $SO_{2\ eq}$	1.36E-01	7.00E-04	2.54E-02	**1.37E-01**	**1.62E-01**
Terrestrial ecotoxicity	kg 1,4DB $_{eq}$	6.28E-05	1.81E-03	3.30E-04	**1.87E-03**	**2.20E-03**
Agri land occupation	m^2/year	2.75E-03	4.20E-03	8.70E-04	**6.95E-03**	**7.82E-03**
Natural land transformation	m^2	5.57E-05	2.05E-05	6.38E-05	**7.62E-05**	**1.40E-04**

Table D-21: Environmental impact contribution of different operation in the process "Mining operations phosphate rock as P_2O_5-TN"

Main Impact category	Unit	Blasting	Material handling	Internal transport
Climate change	kg $CO_{2\ eq}$	9.77E-02	5.96E-02	1.42E-02
Human toxicity	kg 1,4DB $_{eq}$	3.10E-02	6.71E-02	4.78E-03
PM10	kg $PM10_{eq}$	5.29E-02	2.40E-04	4.07E-05
Ozone formation	kg NMVOCs	2.42E-01	7.20E-04	1.10E-04
Ionizing radiation	kg ^{235}U	5.67E-03	1.03E-02	1.07E-03
Terrestrial acidification	kg $SO_{2\ eq}$	1.35E-01	4.50E-04	7.41E-05

Table D-22: Comparison of the environmental impacts of two product systems of wet beneficiated phosphate rock in Tunisia and the USA

Indicator	Unit	USA	Tunisia
Agricultural land occupation	m^2*a	1.59E-02	6.34E-03
Climate Change	kg CO_2 eq	2.54E-01	3.75E-01
Freshwater ecotoxicity	kg 1,4-DB eq	4.85E-03	5.22E-03
Freshwater eutrophication	kg P eq	3.06E-04	1.50E-04
Human toxicity	kg 1,4-DB eq	1.50E-01	2.47E-01
Ionising radiation	kg U^{235} eq	1.01E-01	2.49E-02
Natural land transformation	m^2	3.85E-04	7.62E-05
Particulate matter formation	kg PM10 eq	2.34E-03	7.18E-02
Photochemical oxidant formation	kg NMVOC	6.72E-04	2.43E-01
Terrestrial acidification	kg SO_2 eq	9.65E-04	1.37E-01
Terrestrial ecotoxicity	kg 1,4-DB eq	1.05E-03	1.87E-03
Water depletion	m^3	8.93E-01	9.63E-01